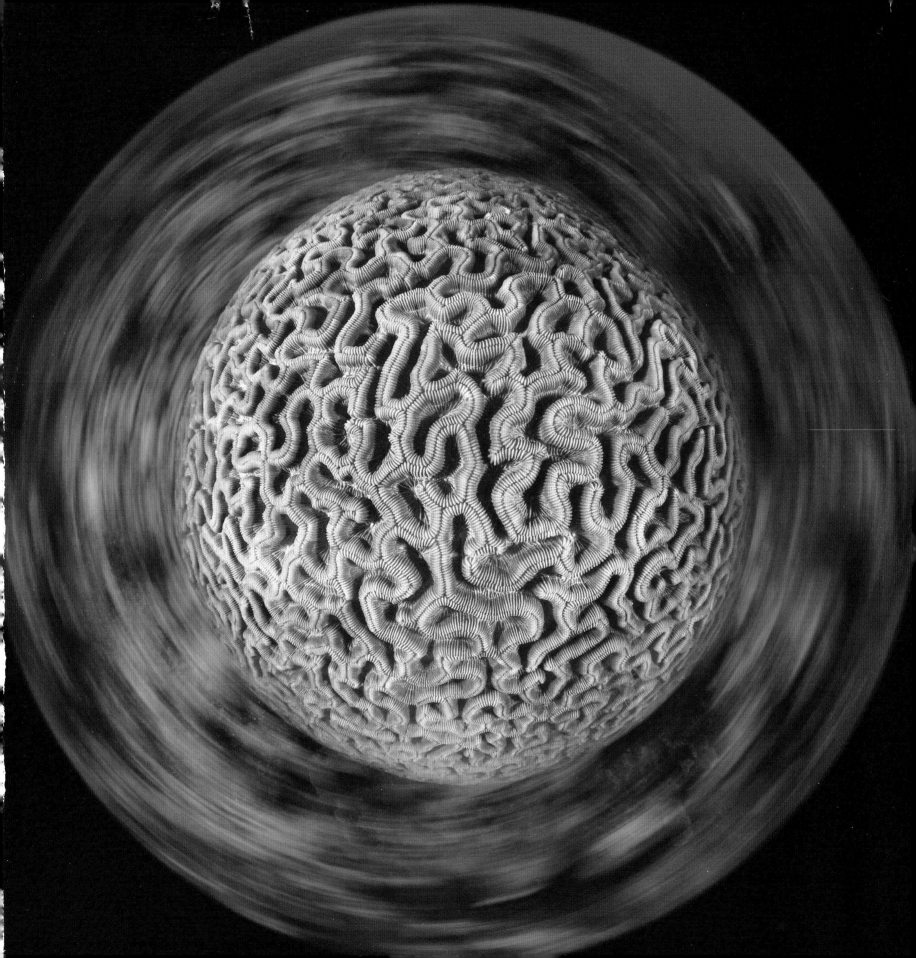

SECRETS

OF THE SEAS

A journey into the heart of the oceans

ALEX MUSTARD AND CALLUM ROBERTS

BLOOMSBURY

LONDON · NEW DELHI · NEW YORK · SYDNEY

Bloomsbury Natural History
An imprint of Bloomsbury Publishing Plc

50 Bedford Square	1385 Broadway
London	New York
WC1B 3DP	NY 10018
UK	USA

www.bloomsbury.com

First published 2016

British Library Cataloguing-in-Publication Data

A catalogue record for this book is available from the British Library.

Library of Congress Cataloguing-in-Publication data has been applied for.

ISBN:	HB:	978-1-4729-2761-3
	ePDF:	978-1-4729-2763-7
	ePub:	978-1-4729-2762-0

2 4 6 8 10 9 7 5 3 1

Designed by Nicola Liddiard, Nimbus Design
Printed in China by RR Donnelley Asia Printing Solutions Limited

To find out more about our authors and books visit www.bloomsbury.com.
Here you will find extracts, author interviews, details of forthcoming events
and the option to sign up for our newsletters.

Contents

Introduction

The sea guards its secrets well. For nearly all of human history, we could only imagine the lives lived beneath the waves. We populated those imagined worlds with gods and nymphs, terrible monsters and endless strange creatures. Or in the cold, dark and crushing pressure of the depths, we saw lifeless voids because it was impossible to conceive anything could exist there. It is only in the last century, especially the last fifty years, that diving equipment and underwater photography have revealed at first hand the real world beneath the sea.

Life began in the sea and has an immensely long marine pedigree. For four billion years, or thereabouts, life has experienced and adapted to huge swings in planetary conditions, waxing and waning as crises came and went. What we see around us today, life with all its beauty, vibrancy, spectacle and secrets, is the product of that long history.

Sophisticated cameras and greater access to remote places allow us to see underwater life in unprecedented detail. This book is a collaboration between a photographer, Alex Mustard, and a marine scientist and conservationist, Callum Roberts. We bring you face to face with creatures from chill northern fjords to the rich heartland of marine biodiversity in the Coral Triangle of southeast-Asia, exploring the past, present and future of ocean life.

The oceans are restless and constantly changing, yet paradoxically appear constant and timeless. That apparent constancy is an illusion brought about by the difficulty of seeing change underwater, and the short blink of time over which most of us know the sea. But today natural and human forces intertwine and human change is gaining the upper hand. The oceans are in a state of rapid flux as a result of human influence, from the coast to the remotest plains of the high seas, and from the surface to the bottom of the deepest ocean trenches. Ocean change challenges everything we think we know about the world. This fact is often ignored or overlooked in photobooks, but not this one. We investigate what change means for the oceans and their inhabitants. While many places have suffered at our hands, we chose not to include pictures of places impacted by human development, greed or carelessness. Instead, we show ocean life in all its magnificence as it should be, and can be again with the right protection.

Sea life is profoundly important to us. This is an ocean planet; most of the space occupied by life is water. But few of us think about it very much. For most people, most of the time, what goes on in the sea is hidden, unseen, unsuspected and ignored. Here we showcase the exquisite variety of adaptations to life in the sea, the interdependence of species and the magnificent spectacles of their lives at scales from the barely visible to ocean going titans. The sea has defined this world since the beginning of creation and its inhabitants are adaptable and resilient. They are responding to the human induced changes in their world in a myriad of fascinating ways which demonstrate their durability and persistence. With a little help from us, the oceans will continue to enthral, inspire and provide as long as there are people to enjoy them.

Riches beyond measure

There is a place where two oceans meet, where warm waters braid and mingle in countless streams that trickle, slosh and flood around more than 27,000 islands. This is the world's largest archipelago, a place where four million square kilometres of shallow tropical sea pulse and thrum with life. Taking in the waters of Indonesia, the Philippines, Malaysia, Papua New Guinea and the Solomon Islands, more different species call this place home than anywhere else in the sea. It is the global heartland of marine biodiversity. Although it covers just 1.5 per cent of the oceans, it supports a third of the world's coral reefs, which is why it is called the 'Coral Triangle'. A staggering 2,500 fish species live here, and more than 600 reef-building corals, three-quarters of all those in existence. These waters blaze with a colour, richness and multiplicity of form that has no parallel.

LEFT A huge shoal of predatory jacks at Tubbataha Reef in the Sulu Sea, Philippines. Unfished reefs are rare today, but isolation and a marine protected area have ensured Tubbataha's reefs remain almost pristine. In places like this we see the strange phenomenon of predators being more abundant than their prey. Usually it is the other way around – think of Lions and antelopes on the African plains – but reefs can sustain a greater weight of predators than prey because prey species are much more prolific and their populations turn over much faster.

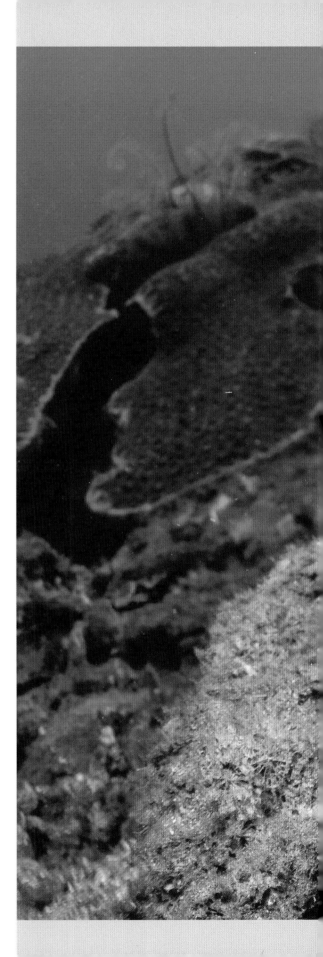

Coral reef fish often surpass the limits of good taste and imagination, like this Paddle-flap Scorpionfish (*Rhinopias eschmeyeri*) in Indonesia. No shape, it seems, is too outlandish, nor colour too gaudy, which is why coral reefs are so beloved of movie makers and animators.

What has gifted the Coral Triangle with such an exuberance of life? It was among these islands that Alfred Russel Wallace hit upon the idea of evolution by natural selection in the 19th century. Like his contemporary, Charles Darwin, he was struck by how isolation on different islands seemed to lead the same species on different paths, each population diverging from others as natural selection built new species from the same clay. Although Wallace spent most of his time in the jungle, he was well aware that the extraordinary diversity he saw continued beneath the sea. A casual stroll along the beach might turn up 100 different kinds of sea shells. The same explanation of evolution by isolation holds one of the keys to the Coral Triangle's incredible marine richness.

For much of the last two million years, the world has been locked in a cycle of freeze and thaw as repeated glaciations gripped the planet. At the peak of each glacial cycle, sea levels fell by up to 130 metres, separating the Coral Triangle into several isolated basins as land bridges emerged between islands. Dropping sea level fragmented species' ranges for tens of thousands of years, enabling them to diverge from one another in isolation. When sea level rise reunited them, there might be several species where once there was one. The repeated fall and rise of the sea made the Coral Triangle evolution's forge, hammering out hundreds of species over vast stretches of time.

LEFT The most striking thing about coral reefs is the sheer head-spinning confusion of fish that swirls above, around and within them. The reefs of Raja Ampat in eastern Indonesia are among the richest of all. A fish spotter could spend months here and still not come close to a complete list. The silver-yellow fish streaming through the middle of the frame are Bigeye Snappers (*Lutjanus lutjanus*), while the blizzard of tiny black and white striped fish in the foreground are juvenile Convict Blennies (*Pholidichthys leucotaenia*). Young Convict Blennies live in deep burrow complexes up to six metres long dug by adults, sometimes more than 1,000 of them to a burrow. This species is one of the few fish on reefs that does not have an open water larval stage, the young instead remaining under the watchful guard of their parents.

BELOW Few animals feed directly on hard corals; their skeletons are too formidable a defence for most teeth and jaws. Bumphead Parrotfish (*Bolbometopon muricatum*), each up to a metre long, like these in Sipadan Island, Malaysia, roam the reef in loose packs and with clashing beaks demolish thickets of branching coral and move on. Such sights are becoming rare, though, as this species is vulnerable to night-time spear-fishing when they rest in groups in shallow lagoons.

ABOVE Coral reefs capture nutrients from the open sea via the 'wall of mouths' – the cloud of plankton-feeding fish – that surrounds them. The fish live in a permanent state of tension, caught between the urge to venture farther from the reef than others so as to be first to reach prize morsels of incoming food, and the risk of becoming food themselves for the roving packs of predatory fish that patrol the reef front.

RIGHT A massive ball of silversides (*Atherinidae*) corralled against a reef wall by pack-hunting Malabar Jacks (*Carangoides malabaricus*) and Kawakawa Tuna (*Euthynnus affinis*).

The second reason for such richness and complexity is so obvious it is often overlooked: this region sustains a greater area of shallow-water habitat than anywhere else in the tropics. Biologists sometimes envy physicists for the elegant simplicity with which they render into a few 'laws' the workings of the universe. The biological world is much messier, they lament, and resists such unifying explanations. But if there are laws in biology, the most fundamental is that bigger areas have more species. They incorporate a wider variety of habitats so more creatures can make a living in them. And because of their greater area they sustain larger populations that are therefore less prone to extinction than in places with less habitat, like oceanic islands.

The Coral Triangle also benefits from geological and oceanographic happenstance, sitting astride the margins of two mighty oceans. It is Grand Central Station to converging currents of both Indian and Pacific Oceans, pulling in species from faraway islands and archipelagos and accumulating them in a bewildering cosmopolitan blend. Swimming through these waters, you are soon overwhelmed by life's variety, dizzied by its abundance. Like a procession of Russian dolls, the closer you look the smaller are the species you see, each one a diminutive but different version of the last, each exquisite in construction and decoration, each perfectly equipped by evolution for the challenges of survival.

Three male anthias, resplendent in their courtship garb. TOP RIGHT is a Squarespot Anthias (*Pseudanthias pleurotaenia*). ABOVE a Purple Anthias, (*Pseudanthias tuka*). BOTTOM RIGHT is a Stocky Anthias, (*Pseudanthias hypseleosoma*). There is a purpose in their showmanship, since male anthias maintain harems of females, mating with them every day throughout long spawning seasons.

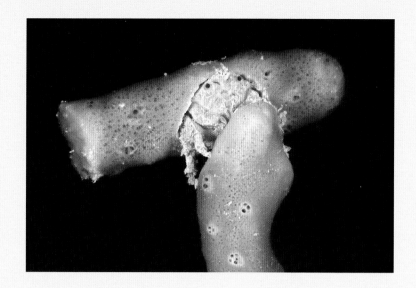

LEFT Coral reefs are defined by the stony corals that build them. By means of an ancient cellular alchemy, corals conjure rock from water to build structures of bewildering intricacy, beguiling beauty and geological endurance. Here a cloud of damselfishes, mostly *Chromis atripectoralis*, plucks plankton from above a tangle of *Acropora* coral in Buyat Bay, Sulawesi. Just as a forest supports a greater variety of species than a grassland, the great architectural complexity imparted by corals helps reefs sustain the extreme richness of life around them.

ABOVE In the fashion conscious world of crabs, this stunning powder blue sponge hat was the seasons big hit. Actually, this sponge crab (*Dromia dormia*) has hijacked a finger of sponge to wear like a hat and make itself inconspicuous and difficult to eat. Few species eat sponges because they are loaded with chemical toxins and microscopic glassy spicules made of silicon.

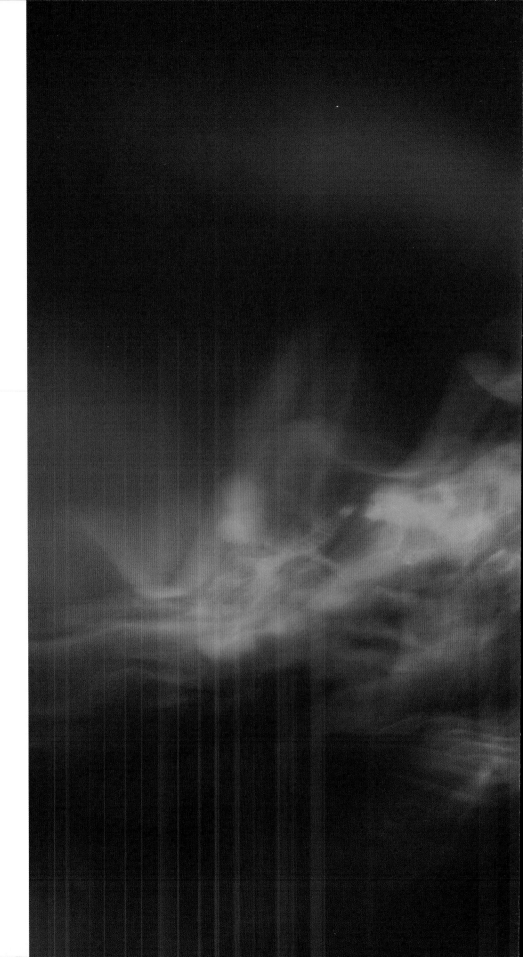

RIGHT A juvenile Harlequin Sweetlips (*Plectorhinchus chaetodonoides*) performs a comical bobbing dance at Negros Island, Philippines. It seems apt that a fish made up as a clown should behave like one, bobbing and nodding, all the while looking at you sidelong as if gauging your reaction to its performance. The dance is probably defensive because it speeds up if you move your finger close to the fish. These fish give up dance when their colourful juvenile pattern changes to the honeycomb attire of adult fish.

This is a world of illusion: football-sized stones covered in seaweed, bright blue sponges, purple seafans, waving fields of grass, all reveal themselves as fish on a second look. It is a place of camouflage and counter-camouflage, of near-perfect concealment, in which animate and inanimate combine with baffling trickery. But other creatures here take a different tack, advertising themselves with outrageous colour combinations, extravagant behaviour or preposterous body shapes. No matter how absurd or outlandish they seem, each is a model of success, surviving against the odds in a sea full of predators.

ABOVE How does a creature arrive at such perfection in concealment? This juvenile red Hairy Frogfish (*Antennarius striatus*) could so easily be mistaken for a rock crusted in sponges and hydroids. Even with close scrutiny, it is hard to place fins, eyes and mouth, to separate fish from fakery; the embodiment of nature as artifice and art.

RIGHT The branches of this red whip coral (*Ctenocella* sp.) are as plush and richly coloured as any theatre curtains, lending this Golden Damselfish (*Amblyglyphidodon aureus*) the air of a player reluctant to make its appearance on stage.

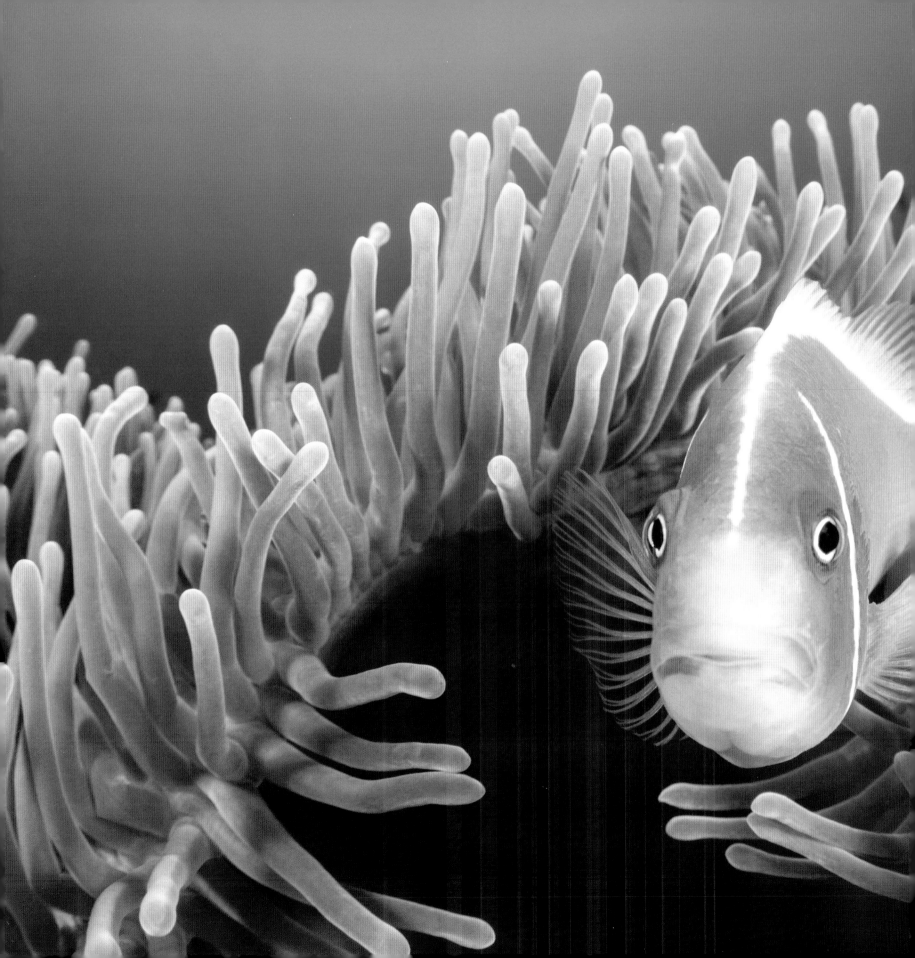

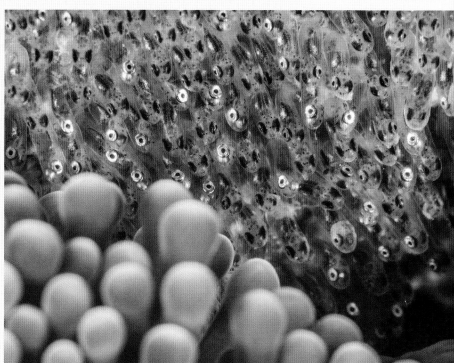

LEFT The Pink Skunk Clownfish (*Amphiprion perideraion*), is a protandrous hermaphrodite, which means that it takes the role of male first in life, then may change to become a female. Each anemone hosts only one female and she is the largest fish present. The breeding male is the next biggest. Any other fish in the anemone are progressively smaller non-breeding males. Being as big as possible maximises the number of eggs a female can lay.

ABOVE Anemonefish eggs develop under the security of the stinging fringe of an anemone in the Philippines. Eggs are rich food and without close care would be stripped in seconds by sharp-eyed predators. Hatching is imminent here, with hundreds of eyes staring from their glassy cocoons. When they hatch, the larvae will swim up and away to escape the battery of mouths on the reef, joining the plankton of the open sea for a few weeks until they have grown large enough (about the size of a small fingernail) to find another reef and seek their own anemone to live in.

Immersed within seas so over-brimming with life, your thoughts begin to drift: how did such wonders come to be; why are there 20 different ways to be a plankton-feeding damselfish; why 500 ways to be a crab? Callum was once asked at the end of a lecture by a member of the audience, "What is the purpose of whales?" It is tempting to see the world in selfish human terms, as if species were only there to serve our needs. Such thinking leads to unfortunate places, to asking for example, how many species do we really need, how many are expendable, which are most valuable? Is a tuna more valuable than a mole crab because we eat one and not the other? Is an angelfish worth more than a cryptic blenny because its size and beauty make it easier to admire? Do you really need 400 coral species to build a reef?

RIGHT Between the devil and the deep blue sea. These Slender Silversides (*Chirostoma attenuatum*) have been driven dangerously close to the reef by a pack of hunting Orange-spotted Jacks (*Carangoides bajad*). Sensing an opportunity, a Slender Grouper (*Anyperodon leucogrammicus*) emerges from the coral, its tail coiled like a spring in readiness to pounce. The ability to exploit fleeting chances like this has been finely honed by evolution.

LEFT There seems no limit to the number of species that can be packed into a coral reef, but a peak is finally reached at Raja Ampat in eastern Indonesia. The Coral Triangle is the global centre of biological diversity and Raja Ampat is its glowing bullseye of maximum richness. Nowhere else in the world has greater diversity and by a combination of isolation and good fortune, the reefs here are still exquisite and intact. They are now the subject of intensive protection efforts to keep them that way for ever.

ABOVE Three Commerson's Frogfish (*Antennarius commersoni*) cling to a vertical reef wall, two of them blending perfectly with the blue barrel sponge nearby. This concealment tactic is both defensive and a means of attack as the frogfish sit and wait, still as the real sponge, until unwary prey approach. This scene may not appear romantic, but most likely the two smaller males in the background are courting the large female in the foreground. She's quite a looker.

In our increasingly human-dominated world, such questions gain sinister undertones. What does it matter if we lose species to human development, pave over coral reefs or cut down forests? With so much diversity around, who could reasonably argue that we cannot do without some of these creatures? This is a slippery route towards poverty. It implies a progressive ratcheting down of the beauty and wonder of the world. In truth, it is impossible to value a sea slug or prawn. What price can you place on the joy and delight such creatures provide? Like the Mona Lisa, Egypt's Pyramids or the Grand Canyon, they are beyond price. They deserve love and protection wherever they occur, to be cherished by those alive today and generations yet to come. ✳

ABOVE A miniscule Emperor Shrimp (*Periclimenes imperator*) only a centimetre long rides shotgun on a much larger T-bar Nudibranch (*Ceratosoma trilobatum*) in the Molucca Sea. The shrimp uses the sea slug as a mobile feeding platform, plucking lumps of mucus from the slug's back and even picking through its faeces as the slug defecates. The sea slug is toxic so predators avoid it, giving its shrimp rider protection too.

RIGHT Coral reefs produce hundreds of new discoveries every year. This strikingly beautiful pair of tiny nudibranchs, each no bigger than a fingernail, were photographed in Seraya Bay, Bali, and are from the species *Doto greenamyeri*, which was newly described in 2015. They feed on the stinging hydroid on which they are resting. Do hydroids tingle on sea slug tongues like spicy chilli peppers do on ours?

What is natural?

In this fast-changing world, we are used to the stories of our grandparents being out of kilter with our personal sense of reality. But there are some places where the stories of old-timers are in such jangling discord with the evidence of our own senses that it is hard to believe they were ever true. The North Atlantic is one such place. To the many of us who live in countries that ring the North Atlantic, it is a place we feel we know well. Memories of seaside holidays are steeped in scents of kelp and brine, bracing murky water, the clash of pebbles and hiss of waves on windswept coasts. It is a region of mists, fleeting sunshine and of seascapes rendered in a chilly palette of greys, greens and browns.

LEFT Atlantic Cod (*Gadus morhua*) gather to spawn in their thousands in Thorshofn, Iceland. Scenes like this were once commonplace all over the North Atlantic from the Baltic to Cape Cod, where cod have supported prolific fisheries for 1,000 years. But stocks collapsed in Canada in the 1990s due to industrial fishing and have declined steeply in Europe where warming seas have made matters worse for them. However, they thrive under good management and favourable environmental conditions in their northern strongholds of Iceland, Norway and Greenland. This picture was taken during a two-week pause in the fishing season to allow some cod to spawn undisturbed.

The North Atlantic is a place of raw power and extraordinary productivity. Winter storms and ripping currents mix nutrients from seabed to surface, priming the sea for an eruption of plankton come springtime. The plankton feed hosts of shellfish, fish and marine mammals that live mostly hidden lives, from time to time revealed by the clamour of seabirds raining attacks on some distant patch of water. Colourful boats bring the fruits of the sea to harbour, landing boxes of scallops, crabs, prawns and a jumbled assortment of bottom fish brought up by the trawl net: flat, round, knobbed, spined, scowling and serene. It seems so timeless and natural, the worlds of people and ocean interweaving in perfect harmony.

ABOVE Atlantic Wolffish (*Anarhichas lupus*) jaws are full of crooked, crystalline peg teeth that are perfect for gripping and crushing their prey of sea urchins, molluscs and other shellfish. They hide in rock fissures and caves, but their presence is given away by the scatter of shell remains nearby.

RIGHT An Atlantic Wolffish (*Anarhichas lupus*) lurks in the shadow of a rugged volcanic cleft in Iceland. These creatures reach up to 1.5 metres long and large individuals can be as thick as a man's thigh. They were once common across much of the North Atlantic, but numbers declined steeply as bottom trawlers dragged their nets ever more widely during the 19th and 20th centuries. Their white, flaky flesh was considered a good substitute for cod, while their tough, highly patterned skins were tanned into leather for belts and purses.

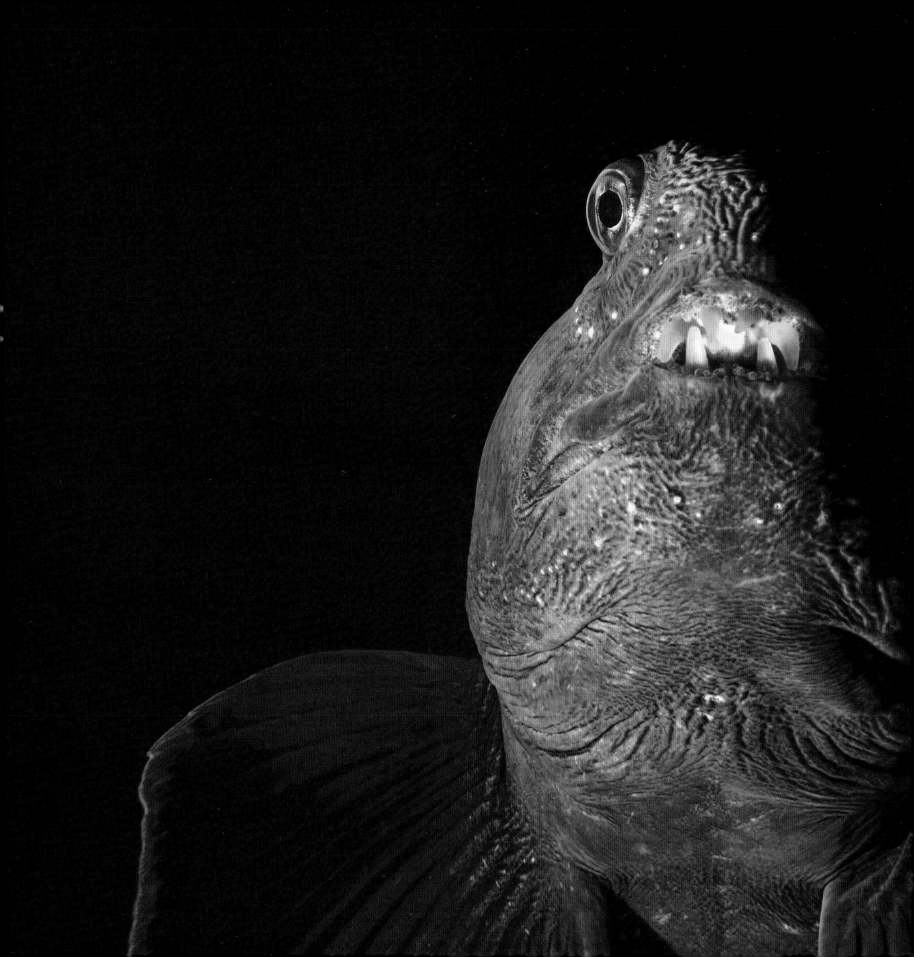

But the scenes of today are very different to a century ago, or even 50 years. The familiar face of the Atlantic has more to do with how we use this place than to natural processes running their course unfettered. Many of the creatures and habitats that once dominated this world have been pushed to its edges, supplanted by those better able to live alongside intensive human activity. Toothsome wolffish, broad and muscular conger eels, angel sharks, Blue Sharks, thronging Spiny Dogfish, bluefin tuna sleek and swift as torpedoes, giant halibut, skates as big as dining tables, seething, roiling shoals of cod, all once held court in these waters. They lived within easy reach, often close inshore. An account of a night's fishing in the North Sea from the UK's Holy Island in 1834 gives a good impression of their former numbers and size:

RIGHT The rising tide conjures a forest from lank seaweed-strewn rocks around the Isle of Coll, northwest Scotland, lifting slippery fronds into glades and arbours frequented by animals like this Grey Seal (*Halichoerus grypus*).

BELOW Kelp can grow astonishingly fast, adding tens of centimetres per day when spring comes to northern seas. They soon fashion a labyrinthine habitat for hundreds of other creatures, like this nudibranch *Polycera quadrilineata*. The canopy is stripped again by autumn tempests, and in remote communities of the North Atlantic the seasonal bounty of storm-tossed kelp was once gathered by farmers as fodder and fertiliser.

[On reaching the grounds]... the line was carefully passed over the side, and as soon as one line was nearly run out, the end of another was fastened to it till the whole four were in the water. The setting of this line occupied about an hour, more or less... We then went directly back to the end first put out, and... the work of slaughter began. At first there was an almost interminable number of haddocks caught; [then] on looking carefully down into the "vasty deep" a huge mass, of indescribable appearance, showed itself gradually approaching the surface (by this time it was broad day light)...and when fairly brought into view turned out to be a very large skate. We continued hauling in the line, catching alternately cod—ling—dog-fish—star-fish—and some flatfish... We rowed away to another line which is always set... with hooks about four times as big... baited with large pieces of haddock...Here we had a renewal of the former scene, only all the fish on this line were large. ...We got a large halibut, which was no sooner in the boat than it nearly took forcible possession of it... Our night's work produced about two hundred haddocks, thirty-nine cod, four skate, three ling, a halibut, and a number of dog-fish.

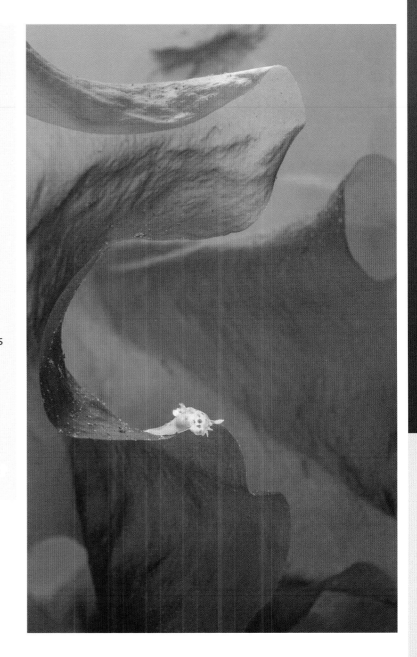

ABOVE Blue Sharks (*Prionace glauca*) hunt in the English Channel. These slender creatures are the quintessential predators of the open sea, gathering wherever shoals of prey like sardines and herring congregate. While still one of the commonest large sharks, this species' fins are prized in China for shark fin soup and it has become the mainstay of Atlantic shark catches as other species have dwindled.

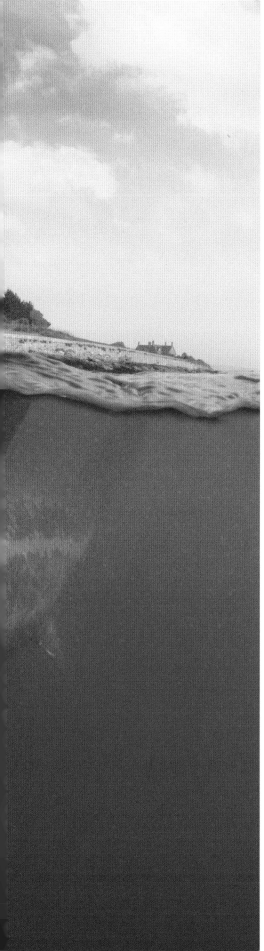

The seabed was different too. Descriptions from the early 19th century are of endless mats of waving invertebrates seen from above through crystal water, oyster-crusted and bejewelled with flashes of colour from sea slug or fish. Today, the beaches of northern Europe are strewn with timeworn oyster shells, some the size of horses' hooves. These ghosts from the past offer silent testimony to lost abundance because where once there were billions, hardly any survive now.

Life in today's North Atlantic has shrunk in numbers and size. The giants of old have yielded to legions of smaller-bodied creatures, like gurnard and sculpin, John Dory and plaice, prawn and clam. The seabed has been cleared by heavy bottom trawls and dredges that work the ground as do plough and harrow on land. Gone is much of the concealing cloak of plant and animal life: today's vistas are of shifting sands and gravel. There is still much beauty and splendour here, especially close up, but this world is as much a human construct as natural.

LEFT Seen close to St. Michael's Mount in the English Channel, this Basking Shark (*Cetorhinus maximus*) surges through water thick with planktonic food so small it is scarcely visible to the naked eye.

ABOVE A Basking Shark (*Cetorhinus maximus*) ploughs through a dense plankton swarm made up mainly of a tiny crustacean copepod, *Calanus finmarchicus*, in the Cairns of Coll, Inner Hebrides, Scotland. Reaching 12 metres long and 4 tonnes, the Basking Shark is the second largest fish in the sea. Large bodies come with huge oxygen demands so the biggest fish species, like Basking Sharks, Whale Sharks and manta rays, eat plankton enabling them to feed and at the same time flush enough water across their gills to breathe.

Most people look upon the North Atlantic as furious and untameable, natural and wild, even those whose responsibility it is to protect nature. But accepting this altered state as natural means we condemn the diminished and disappeared to their exile, rather than helping them return to triumph once more. This chapter offers a vision of past magnificence. All of the vanquished are alive today and still thrive in isolated redoubts that are spared from overfishing by inaccessibility, rugged terrain, good fortune or active protection. The photographs were taken in these remaining strongholds to show the North Atlantic as it once was and could be again.

LEFT Atlantic Bluefin Tuna (*Thunnus thynnus*) are among the most dramatic of the Atlantic's top predators, slashing into schools of prey fish and scattering them in all directions. At one time they streamed into the North Sea and Baltic Sea through the English Channel and around the north of Scotland, in hot pursuit of giant herring shoals. Everything about them is extraordinary. Individual tuna can weigh more than 700 kg, the size of a bull moose, and reach more than 2.5 metres long. Sleek and muscular, they can exceed 60 km/ph and dive a kilometre down. They are also the most expensive fish in the world, with single fish sold in Japan for hundreds of thousands of US dollars. This great value has been their downfall, and intensive fishing has wiped out the North Sea and Baltic Sea stock, and depleted overall numbers by 70 to 80 per cent.

ABOVE If any fish could be said to define the world of which it is a part, it would be Atlantic Cod (*Gadus morhua*) in the North Atlantic, pictured here in Iceland. This animal owes its dominance to rapid growth, large size (up to 1.5 metres), a very broad diet (cod have even been found with ducks in their stomachs) and prolific reproduction. A large female Atlantic Cod can produce more than five million eggs at a single spawning. When he heard this fact, then newly discovered, the 19th century French author Alexandre Dumas mused that if they all survived, within a few years people could walk across the Atlantic dryshod on the backs of cod.

The raw muscular power of a Basking Shark (*Cetorhinus maximus*) as it thrusts its way through rich green waters. These vast animals gather to feed on summer plankton blooms off European coasts and were once known as 'sunfish' for their habit of basking at the surface. They disappear in autumn, which led to a belief that they slept over winter on the deep sea bed. Satellite tags now reveal they undertake long migrations, sometimes crossing the Atlantic, and spend months feeding hundreds of metres below the surface.

RIGHT A female Long-snouted Seahorse (*Hippocampus guttulatus*) in the Canary Islands wafts back and forth with a clump of seaweed. There is something serene about the measured, delicate, unhurried pace of seahorse life.

LEFT A male Lumpfish (*Cyclopterus lumpus*) defends a nest of eggs among shallow rocks in a Norwegian fjord. Also called a Lumpsucker, this fish can attach itself to rocks using a suction pad on its chest made from modified pelvic fins, enabling it to stand guard even in heavy surge.

BELOW LEFT A European Lobster (*Homarus gammarus*), surprised at night in Grevelingenmeer, Netherlands. Lobsters and others invertebrates, like prawns, crabs and scallops, have done well in seas transformed by fishing. The depletion of predators like skates, rays and cod has greatly reduced their risk of being eaten, although not by us!

BELOW RIGHT A Butterfish (*Pholis gunnellus*) seems to pose for its portrait. Butterfish are slender and eel-like, but are not eels at all but gunnels, a group on its own. They are among a handful of fish that can tolerate the extremes of the intertidal zone, hiding in puddles under weed or rocks when the tide recedes.

RIGHT A rare photo of spawning Common Dragonets (*Callionymus lyra*) in Scotland. These fish live on the seabed, but here the larger male lifts the female off the bottom balanced on his pectoral fin. Males must compete strongly with others to attract mates and the process is very demanding. According to one study in Plymouth, males breed for only a season, in either their third, fourth or fifth year of life, then waste away and die.

Unwittingly we have transformed life in North Atlantic seas over the course of 200 years of fishing and development. The life that thrives here today is resilient and adaptable. It flourishes despite us and because of us. But the balance between human dominance and nature has swung too far. We need to reset it so that more sensitive and vulnerable species, like those in the photographs, can recover and prosper. To do that, nature needs a network of refuges placed permanently beyond the reach of trawl and dredge, hook and net. It isn't asking much and would not mean foregoing fish catches. By fishing less, stocks will rebuild so we can catch more from smaller grounds at less expense. That way there can be real harmony between people and nature. ✶

RIGHT Two male Common Cuttlefish (*Sepia officinalis*) compete for a female (on left) during spring courtship in England. Although not visible in this photograph, the middle male adopts a split personality in this encounter, flashing his tiger-striped flank to the male on the left, while showing courtship colours to the female. Cuttlefish can change colour from moment to moment, their skins flashing like pulsing light shows. This remarkable ability comes from pocket-like structures in the skin called chromatophores, which come in three colours that are further modulated by silvery pigments. Chromatophores can expand to show the colours, or contract in a split second, producing waves of colour, bold patterns or perfect camouflage as circumstances demand.

PREVIOUS PAGES Spiny Spider Crabs (*Maja squinado*) gather in their tens of thousands at Burton Bradstock, UK, to moult their shells and mate. Temperate waters are highly variable through the year and often host spectacular seasonal phenomena that take place when conditions are just right. Soon after mating, the crabs disperse to continue their solitary lives among rocks, seaweed and sand.

RIGHT Beneath the seaweed canopy lies a lower understory of shrubby weeds, corallines and invertebrates. For the tiny creatures that live there it is dark, mysterious and dangerous, offering concealment to the most terrifying beasts, like this skeleton shrimp, *Caprella linearis*, in Iceland. But appearance is deceptive here, since the skeleton shrimp feeds mostly on muck, picking and sifting through it with its large forelimbs.

LEFT Eye to eye with a Northern Prawn (*Pandalus montagui*), perched within a colony of Dead Man's Fingers (*Alcyonium digitatum*).

Perfection
in motion

At the edge of visibility in the blank blue of open sea, half seen, half imagined, a shadow moves. Detaching itself from the wavering patterns of light and shade that confound all sense of perspective in this featureless world, it resolves into the shape of a shark, blue-grey on blue, like a figure emerging from mist. Bold and deliberate, it approaches quickly, veering at the last to circle, one eye appraising, considering the strange human form in its water world. Then with an audible tail flick it passes on, arrow-straight, to disappear once again into implacable blue. It leaves in the watcher a frisson of excitement, touched with awe and fear.

LEFT An Oceanic Whitetip Shark (*Carcharhinus longimanus*) cruises in the emptiness of the big blue. It is hard to imagine what life must be like in a realm so devoid of physical references, a place defined more by absence than presence. Like a bird of prey in the open sky, a predator needs speed and power to hunt in the void.

BELOW The Bahamas is one of the most reliable places in the world to come face to face with a Tiger Shark (*Galeocerdo cuvier*). Close encounters like this are softening hostility towards this fish, revealing it as a considered predator, not a mindless killer.

RIGHT The diamond gape of a Basking Shark (*Cetorhinus maximus*) surging through thick plankton, visible as dust-like specks suspended in the water. The pale surface inside the mouth attracts plankton, countering their natural tendency to dart away at the approach of a predator.

Sharks seem primordial. So long ago it is almost impossible to grasp, evolution appears to have perfected the shark. Fossils of fish that are recognisably sharks appear 400 million years ago. Like today's sharks, their bodies were covered in tiny skin denticles, or miniature grooved scales. Denticles reduce drag, allowing sharks to swim faster using less energy. They had lobed tails, torpedo-shaped bodies and pectoral fins shaped like hydroplanes. And lots of sharp teeth of course. Early sharks were small, but evolution soon scaled them up into predators capable of dispatching large prey. Sharks have held to the same basic plan, playing with variations on a theme, through vast geological epochs and recurrent planetary upheavals. They have endured several mass extinctions and survived long stretches when the oceans became more acidic and were starved of oxygen. They thrive best in the few hundred metres near the surface where food is plentiful, and they hit a barrier at around 3,000 metres deep, which their physiology has never allowed them to penetrate.

BELOW LEFT There is something Disneyish about the grin of this Great White (*Carcharodon carcharias*), although it wouldn't seem at all funny if you were a seal pup struggling to perfect your swimming technique.

BELOW RIGHT There is something ominous about the classic chevron profile of a shark approaching head on. Here an Oceanic Whitetip (*Carcharhinus longimanus*) makes a beeline for Alex. Unlike many other sharks, which circle novel objects like divers before coming in for a closer look, Oceanic Whitetips speed directly toward you, intent on investigating any object in their monotonous world. They will often make an exploratory and somewhat unnerving bump before moving off.

Sometime between 230 and 200 million years ago, evolution spawned a new way to be a shark, creating a lineage that abandoned the open sea for life on the bottom. Stingrays subsequently blossomed into hundreds of species, filling niches all across shallow seas. In turn, they lifted back off the seabed into open water as plankton-feeding devil rays about 30 million years ago, from which manta rays split about 10 million years later.

How long have sharks haunted the human imagination? They are consummate predators in a world where we are ungainly and very recent visitors. Their approach is silent, swift, often unseen, and for those marked out as prey, almost invariably lethal. We have a long aquatic legacy that reaches back more than 150,000 years to times when our ancestors first gathered seafood from shores exposed at low tide. We would have seen their dorsal fins draw criss-cross ripples along the surface, watched from cliffs as broad-backed giants hunted, might have marvelled as they churned the sea to foam tearing into shoals of fish. Shark and tuna bones found in a Timor-Leste cave once occupied by humans show that by 43,000 years ago, the hunters had become the hunted.

LEFT Although out in the open with no cover, these courting Tassled Wobbegongs (*Eucrossorhinus dasypogon*) blend almost perfectly with the coral plate beneath.

BELOW Some sharks spend most of their lives motionless on the bottom, waiting for food to come to them. This Tassled Wobbegong (*Eucrossorhinus dasypogon*) in West Papua, Indonesia, is an ambush predator. Like all sharks and rays it has electro-sensitive pores, especially around the face and mouth. These help it detect prey fish when they come within range. The backward-curving teeth ensure the journey from mouth to stomach is one way only.

LEFT A Tiger Shark (*Galeocerdo cuvier*) breaks the surface at dusk on Little Bahama Bank. Tiger Sharks were once among the most common of large sharks, and most feared, but like Tigers on land, they have declined in the face of human persecution. There are no global estimates of their population size, but wherever sharks have been fished, Tiger Shark numbers have fallen.

ABOVE Great White Sharks (*Carcharodon carcharias*) look very different from above. Their sleek torpedo profile and the slender, tapered connection between body and tail are transformed into the thickset stockiness of a prize bull. The tail is no more than a flat blade, powered by thick muscles to either side.

Evolution may have created a great survivor, but it didn't prepare sharks for industrial fishing. How it happened is lost to history, but the gelatinous noodle-shaped structures in shark fins are considered a delicacy by Asian cultures, and eating shark-fin soup has become an expensive symbol of prestige. It is hard to calculate the scale of the killing when many sharks simply have their fins hacked off at sea and their bodies are never landed. Our best estimate puts the toll at 100 million a year. In the last 30 years or so, there have been precipitous declines in abundance of many species, including the majestic Oceanic Whitetip. Once one of the most abundant large animals in the sea, the Oceanic Whitetip has declined 1,000-fold across much of the sea, and can now only reliably be encountered by divers in two places: the Bahamas and remote parts of the Red Sea. As abundant sharks have declined, so fisheries have shifted to target others, including animals like mantas and devil rays, endangering them in turn.

ABOVE You might think that the biggest fish in the sea would be the best known. But the life of the Whale Shark (*Rhincodon typus*) remains mysterious. The ones we encounter near coasts are usually juveniles, like this five- or six-metre fish off Cancun, Mexico. Adult Whale Sharks reach 12 to 15 metres long and spend most of their time offshore. Position- and depth-sensing tags attached to sharks reveal that they can migrate thousands of miles between rich patches of food, and may dive down to 1,000 metres to reach dense concentrations of plankton. Deep dives have to be spaced with long periods at the surface to warm up and recharge with oxygen, since deep water is cold and has less oxygen than that at the surface.

BELOW Two grizzled male Great White Sharks (*Carcharodon carcharias*) power past near Guadalupe Island in the eastern Pacific. The one in the background has been fitted with a transmitting tag to beam data on its movements to scientists. Tagged sharks here make long seasonal migrations across the Pacific to a mysterious spot in the middle of nowhere that has been dubbed the 'White Shark Café'. They linger there, making repeated bounce dives into deep, chill water, warming up near the surface in the intervals. Nobody is sure whether they go here to feed or breed, or perhaps both. But Bigeye Tuna (*Thunnus obesus*), a potential prey for the sharks, aggregate here at the same time, so feeding is a strong possibility. Some sharks vacation in Hawaii on their round trips from the Mexican coast.

LEFT Southern Stingrays (*Dasyatis americana*) pass like shadows through the Caribbean Sea at dawn. It is a common misconception that sandy seabeds are flat and featureless. As well as the ridges created by water movement, large animals like stingrays dig pits and furrows when foraging that add complexity, offering shelter and opportunities for many other creatures that live in these open habitats. European seas once abounded with giant skates much like these stingrays, but they gradually disappeared as fishing intensified.

Sharks once played a dominant role as top predators in the oceans. We are only now coming to understand just how important this role was as the sharks disappear, producing unanticipated upsets in the balance of life. Loss of a slew of big sharks on the US Atlantic coast has destroyed fisheries for Bay Scallops because a scallop predator, the Cownose Ray, has exploded in numbers after its release from control by sharks.

Even as sharks have declined, our perceptions of them are mellowing. The hit 1970s movie *Jaws* played on our primal fear of being eaten and signalled a nadir in people's attitudes. But today sharks are more admired and respected than feared. A recent incident in Cape Cod, close to where *Jaws* was filmed, shows the shift in attitudes. A beached Great White Shark was saved from death and released back to the sea by beachgoers and coastguard.

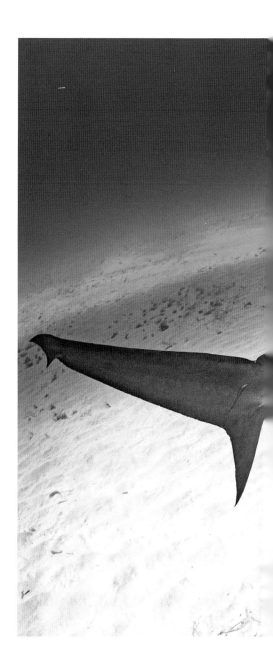

BELOW Great Hammerhead Sharks (*Sphyrna mokarran*) are among the biggest predatory sharks, with females like this one occasionally topping six metres long. When approaching underwater their burly chests are as wide as the fully stretched arm span of a large man. Macho fishermen (they are usually men!) have pitted themselves against great hammerheads for many years, especially off the coast of Florida, catching mainly pregnant mothers that migrate long distances to give birth there. Controversially, this fishery continues today, despite a rapid fall in shark numbers along the US east coast.

RIGHT The Whale Shark (*Rhincodon typus*), here viewed with a fisheye lens, can weigh more than 20 tonnes. There are at least four in this photo, drawn to this spot off the Caribbean coast of Mexico by the abundant planktonic food. Spots hold the key to their identities. Scientists adapted a pattern-recognition computer program, used to identify stars from images of the night sky, to identify individual sharks from photographs of the unique constellations of spots on their flanks.

RIGHT Blue Sharks (*Prionace glauca*) are ocean wanderers, moving hundreds or thousands of kilometres from place to place to find concentrated food patches, like shoals of Mackerel or Herring. These fish were in the English Channel off the coast of Cornwall. In centuries past, Blue Sharks were seen here frequently during the pilchard season when vast shoals of Pilchards came close inshore to spawn. Overfishing of the Pilchards left the sharks with little to eat so they moved off in pursuit of other food. Perhaps the most surprising thing about this shot was the fact that the water was 20°C at the time. As the sea warms, climate change is leading to poleward shifts in the ranges of thousands of species all over the world, including the mackerel these Blue Sharks eat.

RIGHT Storm clouds lend an air of menace to this female Southern Stingray (*Dasyatis americana*), but she is harmless unless provoked. Captain John Smith, founding member of the 17th century colony of Jamestown in America's Chesapeake Bay, speared one with his rapier and swiftly regretted the act as he was stabbed back by its poisonous barb. In the agonising aftermath, he chose the site for his own grave but fortunately didn't need to use it, recovering a couple of days later.

BELOW With its wingspan stretching up to seven metres, the Giant Manta Ray (*Manta birostris*), here with suckered remora (*Remora remora*), is the largest of all rays. Until recently these giants' lives have mostly been untroubled by people, although in the early 20th century there was a sport fishery off the east coast of the United States in which 'gallant' hunters pursued them with harpoons. Today, however, they are hunted for their gill rakers, the apparatus used to sieve planktonic food from the water, which have gained currency in Asian traditional medicine. But Giant Mantas are ill-suited to being fished as they reproduce infrequently and produce only one offspring at a time. So great is the concern that they have been added to the Convention on International Trade in Endangered Species in an effort to protect them.

Today, people travel thousands of miles to dive with sharks, relishing encounters even with species that once provoked terror, like Tiger Sharks and Great Whites. We are mesmerised by their elegance and mastery. Sharks ride stiff currents as eagles do on updrafts, effortlessly holding steady or swooping away with a scarcely perceptible tilt of the fins. On Palau, a live shark is estimated to be worth 1,000 times more than a dead one because of its value to the tourist industry. Whether this reappraisal will be enough to turn shark fortunes remains to be seen. Certainly, vigorous efforts are now underway to protect them with nations like Palau and the Bahamas declaring their waters 'shark sanctuaries' and California banning the sale of fins. While these efforts might still seem inadequate to the scale of the threat, there is hope from history. Sea turtles were in the same dire spot in the 1960s and 70s due to overhunting for their meat and shells. But campaigners shifted perceptions and turtle soup and shells fell out of fashion. Today, most are well on the road to recovery. ✷

FAR LEFT Hammerhead sharks (here a Great Hammerhead, *Sphyrna mokarran*) have mystified us for hundreds of years. Their peculiar heads with eyes set at the end of flat bladed stalks are perplexing. What on earth is the function of such a strange anatomy? Theories abound, such as the idea that by setting their electro-sensory pores wide apart they can sense with them in stereo. However, another use has been spotted. A Great Hammerhead hunting stingrays on the Bahama Bank ran down a ray, pinned it to the seabed with its flattened head and then while holding it to the bottom, manoeuvred for a bite from one wing. It then repeated the process to disable the ray before dispatching the rest of the fish half an hour later.

LEFT Atlantic Devil Rays (*Mobula hypostoma*) fly in formation off the coast of Cancun, Mexico. Devil rays are not at all demonic. They are named for the head fins that funnel prey into their mouths, which when not in use and rolled up, look like horns. Once considered worthless, like mantas they are now hunted for just their gill rakers, an ingredient in Chinese medicine. As these fish produce only one pup at a time, fishing will cause rapid decline so a campaign is underway to protect them.

A Spotted Eagle Ray (*Aetobatis narinari*) digs for prey among seagrass in the shallow lagoon of Grand Cayman Island in the Caribbean. These rays use their shovel-shaped snouts to excavate clams, urchins and other shellfish which are then crushed by ridges of flattened teeth. Feeding rays are often accompanied by mobs of small fish and even cormorants, which wait to snap up animals disturbed or dug up from the sand.

Transitions

"The edge of the sea is a strange and beautiful place," wrote Rachel Carson, the American environmentalist and marine biologist in 1955. It has always been a place of transition: from liquid to dry land, salt to fresh, and buoyant weightlessness to gravity's heavy pull. Some creatures have made the journey across this boundary during evolution's long haul, while others live on the margins, or split their lives between different worlds. This chapter celebrates such transitions.

LEFT Atlantic Puffins (*Fratercula arctica*) spend most of their lives at sea and are excellent underwater swimmers, which is how they catch small fish. During the breeding season they load their bills with many slim silver bodies held sideways, before returning to their nest. Their wing-powered swimming is rather robotic, with jerky flaps and changes of direction. Bubbles pressed from their feathers by the water leave a trail that lingers. After breeding, puffins temporarily lose the power of flight as they moult their primary feathers, floating at sea in straggling groups.

Far away in the lost worlds of deep time, 370 million years ago or thereabouts, a lobe-finned fish, perhaps somewhat like today's coelacanth, lived at the edge of the sea. Over hundreds of thousands of years, it made tentative forays into the pools and creeks of deltas and estuaries, heaving itself across sand bars and banks when the need arose. As eons passed, creatures adapted so well to this marginal landscape that they spent longer and longer above water, until a time came when a fish-like creature existed that lived only on land. The first terrestrial vertebrate, or backboned animal, had arrived. For millions of years, the progeny of this evolutionary invention prospered, radiating into many different groups, including the dinosaurs. Life never stands still though, and having conquered land a time came when water beckoned again.

ABOVE West Indian Manatees, or sea cows (*Trichechus manatus*), range throughout the wider Caribbean grazing on shallow seagrass meadows in scattered groups. In Florida, when winter winds nip the Gulf of Mexico, they follow freshwater streams inland to the spring heads, basking in water that never drops below 20°C. When they first arrive they are cloaked in a thick summer fuzz of algae that has grown over their skin at sea. This one has nearly been picked clean by freshwater fish.

BELOW The first European explorers of the New World found West Indian Manatees (*Trichechus manatus*) to be far more abundant than they are today, in places forming vast herds thousands strong. Their flesh was considered excellent eating and pirates and privateers engaged Native Americans from the Mosquito Coast as hunters to keep them well supplied. The manatees in the photograph spend the cold winter months resting in the warmth of Florida's Crystal River, living off their fat reserves. Numbers are slowly recovering following strict protection but the species is still at risk from boat strikes and poisonous algal blooms, or red tides, promoted by pollution runoff.

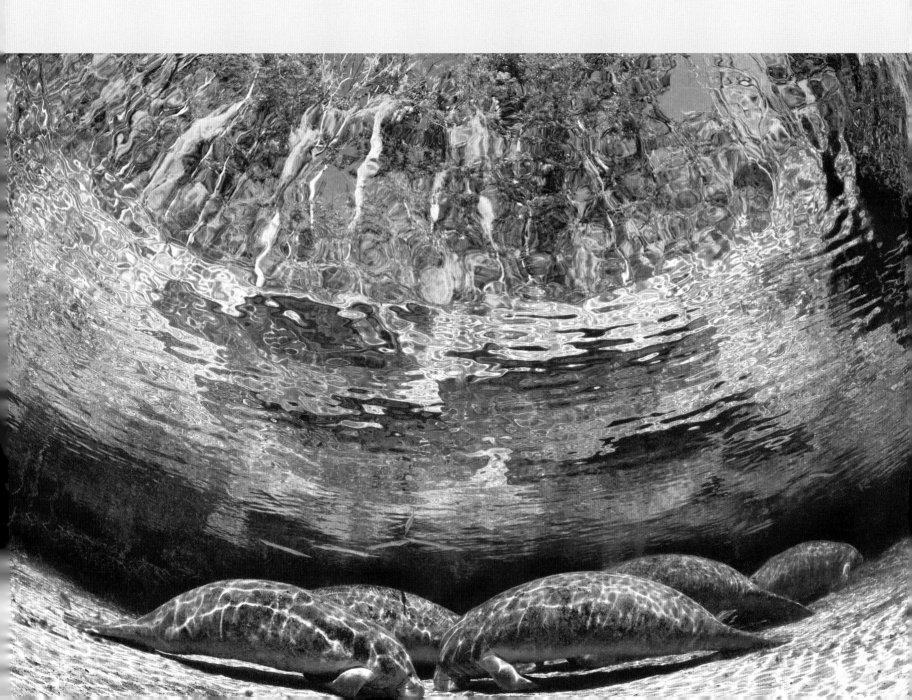

First the turtles evolved around 220 million years ago, moving through freshwater to the sea. Snakes were the last group of reptiles to emerge, around 135 million years ago, colonising the sea sometime after that. Birds that could be called seabirds came on to the scene about 100 million years ago. Mammals appeared earlier, around 205 million years ago, but colonised the sea much later, on at least seven separate occasions as whales and dolphins, sea cows, seals and sealions, Sea Otters, Polar Bears and two now extinct groups, aquatic sloths and the herbivorous desmostylians. Sea cows and whales colonised the sea about 50 million years ago and seals and sealions 30 million years ago.

RIGHT A Green Turtle (*Chelonia mydas*) passes two batfish as it descends from a breath in Sabah, Malaysia. Green Turtles were the main ingredient of 'turtle soup', a once popular dish made not from the meat, but from a gelatinous yellow fatty substance found underneath the lower shell. Turtle soup was served at US Presidential banquets and was a favourite of Winston Churchill, but its popularity led to a steep decline in wild turtle numbers. The dish fell out of favour as conservation efforts grew in the 1960s, giving hope that today a similar change in attitudes and tastes will take shark-fin soup off the menu and save endangered sharks from disappearance in the nick of time.

FAR LEFT A large male Hawksbill Turtle (*Eretmochelys imbricata*) in 'flight'. The demands of swimming and flying are not so different, give or take the huge difference in density of air and water. So turtle flippers work much like bird wings in the air, propelling the animal forwards with broad flaps.

LEFT A Banded Sea Krait (*Laticauda colubrina*) dives back to the reef after taking a breath at Apo Island in the Philippines. These sea snakes are said to specialise in hunting for eels deep in reef crevices, where they immobilise them with fast-acting venom, but Alex has seen them stalking many kinds of fish. When successful, they return to land to digest their prey, lest they become prey for something else underwater while weighed down by their huge meal.

Despite the immense stretches of time involved, marine vertebrates are all still bound in some way to the land they came from. It may be very hard to evolve gills so all are compelled to breathe air at the surface. (Alternatively, perhaps breathing the much richer oxygen in air is a lot easier than extracting it from water, so evolution has been content to retain lungs.) Animals like turtles, sea snakes and birds are still bound to the land by the need to lay eggs, and sea snakes, sea cows and birds must drink freshwater from rivers and streams. The land has been colonised from the sea by other groups, like crabs, but there the ties are in the opposite direction. Even crabs that are fully terrestrial as adults, like the Coconut Crab or other hermit crabs, must return to the sea to launch their eggs on the same planktonic journeys that marine crabs make.

ABOVE While seabirds like terns rely on predatory fish to drive schools of prey fish close to the surface, where they can be plucked from the water, cormorants like this Brandt's Cormorant (*Phalacrocorax penicillatus*) under an offshore oil rig can power their way deep below the surface with their broad paddle feet. They often find themselves hunting amid frantic shoals of prey alongside seals, sharks and tuna. The cormorants are astonishingly nimble. Alex watched one overtake a sealion to steal the fish it was chasing. Surprisingly, the bigger predators ignore the cormorants in favour of fish even though they could easily take them. Fish, it seems, are much better food.

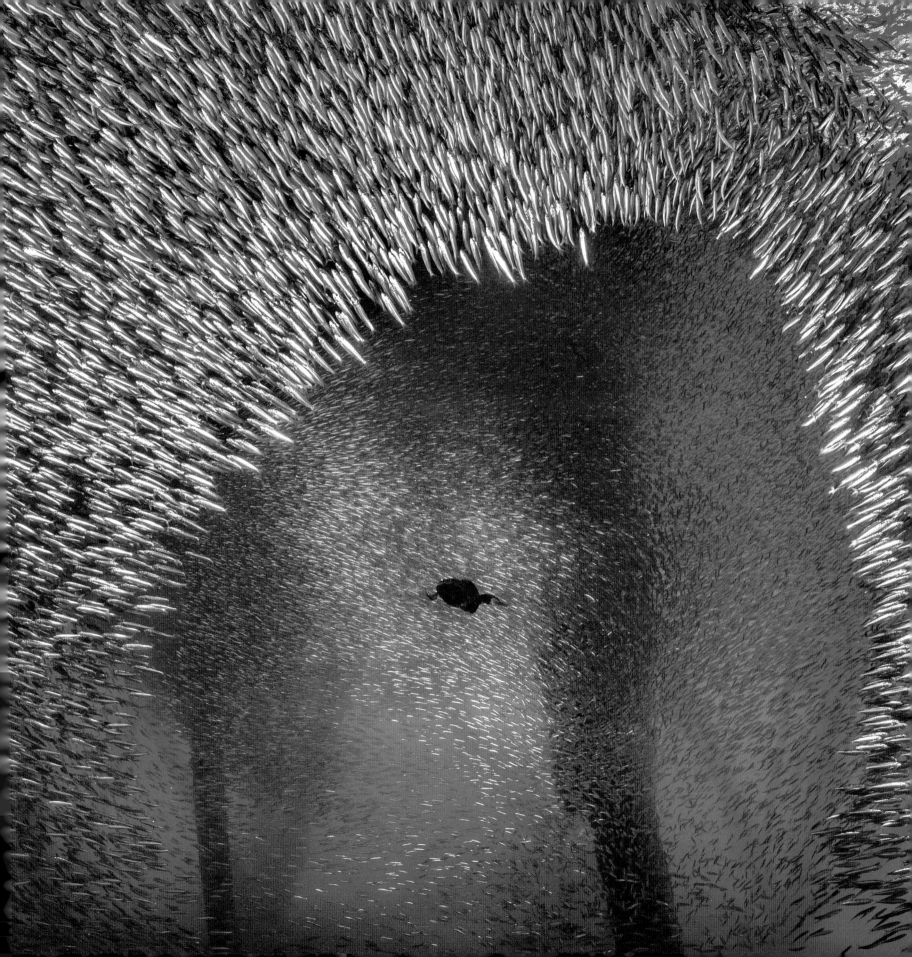

Flowering plants too have colonised the sea having first made the leap from ocean to land as algae some 450 million years ago. It was about 100 million years ago that seagrasses and mangrove trees colonised the fringes of the sea, but they have never made it further than that being tied to shallow places with light to grow and sediment to root. There are no free-drifting flowers in the sea, only seaweeds and single celled phytoplankton ('plant plankton').

While fish have colonised freshwaters from the sea, some fish are physiological chameleons, splitting their lives between freshwater and salt. Salmon live as adults in the sea but have to migrate up rivers and streams to lay eggs. The change in salinity requires a complex bodily adjustment that takes place mainly in the brackish waters of estuaries. Freshwater eels, by contrast, live in rivers and streams as adults, but migrate thousands of miles to spawn in faraway oceans.

LEFT Sockeye Salmon (*Oncorhynchus nerka*) fight their way upstream in autumn to the place of their birth. They spawn just once there and then die, their bodies floating downstream after the final act, to be caught in eddies or washed up on gravel banks. Alex had to wade through rotting bodies to photograph these. The overwhelming pungent, fetid sweetness of decay is something he will never forget!

BELOW Salmon split their lives between freshwater and ocean. This male Sockeye Salmon (*Oncorhynchus nerka*) last fed in the sea weeks ago and its jaw has since grown into an exaggerated curve, designed to compete with other males for the right to sire young with females on the gravel beds where eggs are laid.

Perhaps we are transitional too, but our colonisation of the sea is made possible not by evolution but by ingenuity and technological advance. Seafaring technology goes back much deeper in time than you might think, because we have long lived close to the coast, finding abundant food there. The first known seafood dinners were of shellfish gathered on the coast of South Africa 164,000 years ago and eaten in nearby caves. When modern humans left Africa, they spread fast along the southern coasts of Arabia and Asia, reaching the islands of Indonesia about 60,000 years ago. By the time they colonised Australia, about 50,000 years ago, they probably had rudimentary boats – good enough to fish close to the coast.

ABOVE Taking advantage of a calm in the usually rough sea, Galápagos Green Turtles (*Chelonia mydas agassizii*) graze a luxurious mat of seaweed growing on lava. They share this seaweed with another reptile that has made the jump from land back to the sea, the Marine Iguana (*Amblyrhynchus cristatus*). While turtles make short forays back to land to breed, iguanas make short forays into the sea to feed.

RIGHT Mangrove forests fringe sheltered tropical coasts, extending north and south to places where winter frosts stop them in their tracks. Below their canopy tangled roots trap mud, building up deep layers of peat over thousands of years. Mangroves help protect people and crops on low-lying coasts from tropical storms and it is hoped they can offset the worst of sea level rise from global warming too. By trapping mud, they could enable coasts to grow upwards with the rising sea, but only in places where the supply of mud is sufficient and the trees are not felled for timber or to make way for fish ponds.

As long ago as this sounds, there may have been a closer connection to water in human evolution, perhaps the sea itself. We all carry about with us certain adaptations that make little sense except as a means of better living in and around water. The diving reflex, for example, slows the heart when we plunge our faces in water. We have more fat below our skin than most land mammals, with a body fat content rivalling that of today's Fin Whale. Anthropologists speculate that some ancestor of ours may have paddled in the margins of the sea, swimming and diving for shellfish. Maybe this is why people feel a deep emotional connection to the sea. We love to live near the coast and those who do, according to one study in the UK, are fitter on average and live longer than those inland.

ABOVE Red soft corals (*Dendronephthya* sp.) grow on a root beneath the canopy of a mangrove forest in Indonesia. Although mangrove trees are normally associated with muddy, turbid waters, like those found where tropical creeks and rivers flow into the sea, they sometimes grow close to coral reefs, sharing the brilliant, clear water in which reefs thrive.

RIGHT The Spotted Handfish (*Brachionichthys hirsutus*) is known only from a single place in the world, the Derwent River estuary of Hobart, Tasmania. These fish inhabit the brackish in-between world that separates river from sea. Their pectoral fins are modified into 'arms', with 'fingers' that they use to clamber about on the seabed, hence the common name. Such a tiny geographic range renders a species at high risk of global extinction, should things go wrong. Things went wrong for this creature when Japanese Common Starfish (*Asterias amurensis*) were inadvertently introduced by cargo ships from the Far East. The starfish eat handfish eggs, which are laid on the bottom, putting the handfish in grave danger of complete disappearance.

For much of history the sea has shaped our lives, as barrier, or route to conquest, fame, fortune, flight or exile, a source of food, or of terror from tempest, flood and tsunami. Today it looms larger than ever. The oceans are highways for commerce on which boats carry 90 per cent of goods traded internationally. But more than this, we are coming to understand belatedly, the oceans are pivotal to life on Earth, above and below water. They occupy more than 95 per cent of the living space on the planet and so are of paramount importance in the processes that make the world habitable. That means that what goes on in the sea is not just a matter of idle curiosity but is of central importance to everyone's wellbeing. We have long taken the sea for granted, taking whatever we want from it and dumping into it things we don't want. But we cannot carry on like this. Today, humanity has gained the power to change the planet. We must learn how to exercise that power protecting it. *

LEFT Spinner Dolphins (*Stenella longirostris*) are perhaps the most exuberant of their tribe, performing balletic pirouettes and athletic high jumps as they travel in pods sometimes hundreds or thousands strong. Their solution to the problem of how to corral prey in the featureless emptiness of the open sea is to hunt co-operatively. Groups of dolphins herd prey fish into a tight ball to confine them, while one or two dolphins at a time charge into the mass to gorge at will.

ABOVE All but three of the hundreds of species of blenny in the world live wholly in the sea. But this male Freshwater Blenny (*Salaria fluviatilis*) lives on the bed of a mountain river 700 metres above the Mediterranean in Sardinia, Italy. This blenny and two close relatives, one of which lives in the Greek mountains, all seem to have sprung from a single species that successfully made the leap into freshwater sometime in the past. This ancestor probably possessed the ability to live in fresh and salt water, enabling it to colonise rivers and lakes all around the Mediterranean basin.

Spineless

One of life's great marvels is how living creatures can build monumental habitats from water, air and soil; trees create forests, reeds great swamps. But sometimes the alchemy is even more astonishing, such as when tiny animals build mountains. In the middle of the Namib Desert in West Africa, there is a place so thirsty that hardly a leaf of vegetation softens the time-worn rocks. Low hills rise from a flat plain, rounded by more than half a billion years of erosion. They seem unremarkable, but these hills were made by one of the first reef-building animals in history.

LEFT A Harlequin Crab (*Lissocarcinus laevis*) shelters under the protective spreading bower formed by the stinging tentacles of a tube anemone, *Cerianthus* sp..

Cloudina was a tiny creature, just millimetres wide and up to 15 centimetres long. Nobody is quite sure whether it was a worm or a mollusc, or something altogether different. What makes it significant is that this animal had evolved the ability to extract from seawater calcium carbonate – the basic ingredient of chalk – to build a shelly skeleton. *Cloudina* was in the vanguard of an extraordinary wave of evolutionary innovation that would burst across the planet a few tens of millions of years later, in what is known as the Cambrian explosion of life. Aside from the hills built by *Cloudina*, prior to the onset of the Cambrian 542 million years ago, there are only enigmatic, shadowy fossils in the rocks because the creatures that made them were mostly small and soft bodied. After the Cambrian explosion, recognisable fossils appear everywhere, as if some deity had conjured life itself into being.

ABOVE A female Red-spotted Guard Crab (*Trapezia tigrina*) in defensive posture on a coral (*Pocillopora* sp.). These tiny crabs gain protection from the castellated lattice of coral branches and in return defend the coral against predatory Crown-of-Thorns Starfish (*Acanthaster planci*) attacks, nipping at the starfish's tube feet until it moves on. Crabs also help clear the coral of sediment that would otherwise block light and clog polyps.

RIGHT The great innovation that paved the way for today's coral reefs was the marrying of a photosynthetic microbe – zooxanthellae – with an animal that could precipitate calcium carbonate from seawater. Zooxanthellae multiply coral growth rates to levels where they can lay down carbonate, the basic ingredient of chalk, far more rapidly than it is dissolved or eroded. Corals have an Achilles heel though, a susceptibility to the rising ocean acidity that accompanies higher atmospheric carbon dioxide levels. There are serious concerns that unless we soon abandon fossil fuels, the present era of vigorous coral reef building may be brought to an abrupt end.

What made this a watershed in the history of life was the evolution of hard body parts that fossilise easily. Invertebrates – a catch all name for creatures without backbones – had gained shells and skeletons. Many scientists believe the emergence of predators triggered this revolution. The arms race between predators and prey led to an exceptional flourish of evolutionary creativity as wave after wave of novel beasts spread throughout the global oceans. The age of invertebrates had arrived.

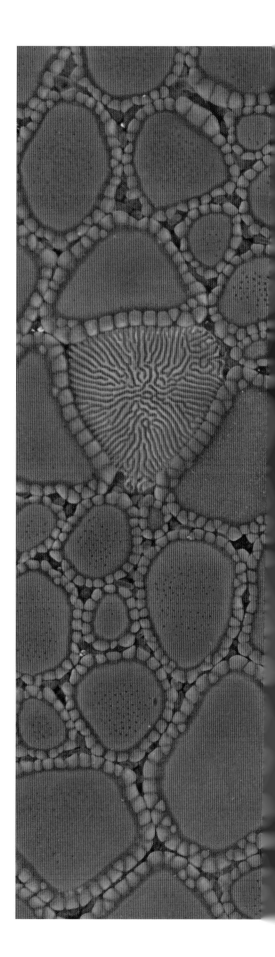

RIGHT The vivid micro-landscape of a Vermilion Starfish (*Mediaster aequalis*). The area shown is about two centimetres wide. The triangular structure that looks like a brain coral is called a madreporite (named for its resemblance to coral – corals were once called madrepores). The madreporite is a valve that connects the radial, internal seawater circulation of the starfish with the sea outside its body. Tiny beating, hair-like cilia pump water in through the madreporite to pressurise the system, powering the tube feet starfish use to walk about, find food and attack prey.

FAR RIGHT Science can reveal unexpected wonders. This colourful aggregation of tiny isopods, *Santia* sp., a creature related to the woodlouse, on the surface of a blue sponge, *Haliclona* sp., looks striking but there is so much more to it than meets the eye. The colours come from tiny unicellular, photosynthetic microbes that they 'farm' on their bodies. The isopods gather in open sunlight so the microbes can grow. They would surely be snapped up by predators but the microbes provide a highly unpalatable chemical defensive shield. The isopods consider them tasty, plucking the microbes from their bodies like succulent summer-ripened oranges (bare patches are where they have scraped them off). The two colour varieties derive from differently coloured microbes; different species perhaps? There are also vanishingly small copepods on the sponge. Who knows what secrets they hold?

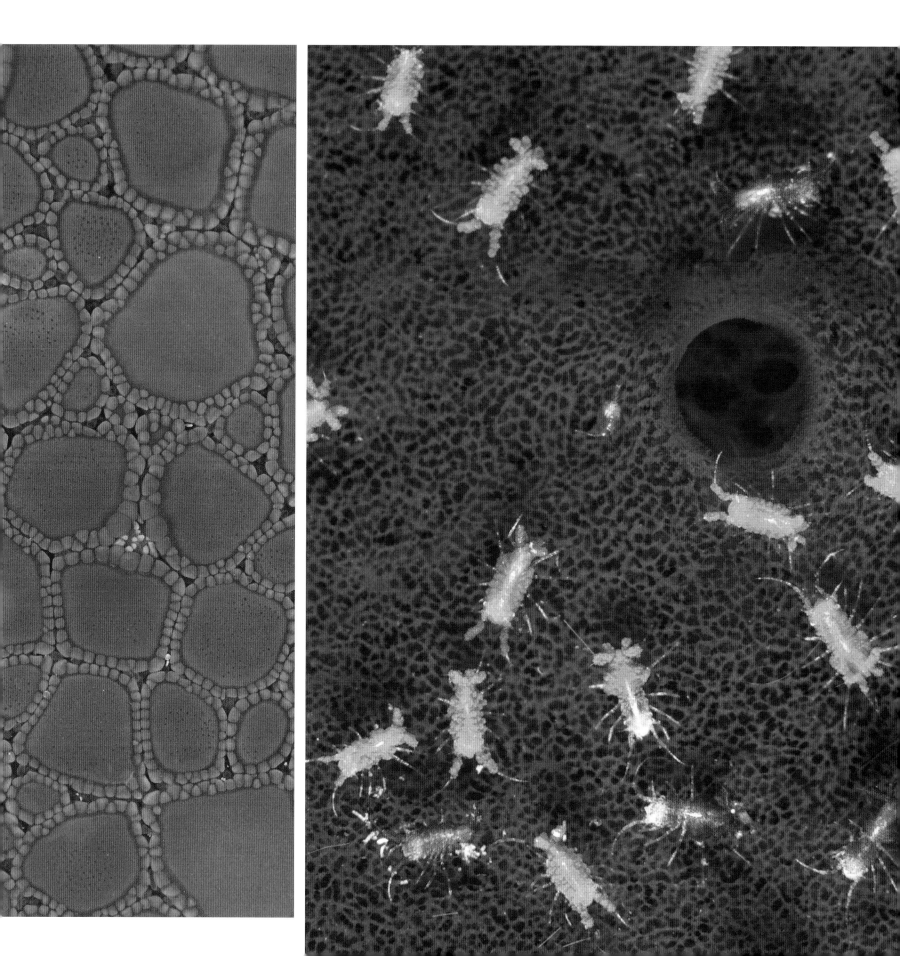

RIGHT Life on a coral reef is gorgeous and surprising at whatever scale you look at it. Zoom in closer and new worlds are revealed. Here a minute pair of Spiny Tiger Shrimps (*Phyllognathia ceratophthalmus*) hold court on top of a coral.

LEFT A Harlequin Shrimp (*Hymenocera elegans*) on a Blue Sea Star (*Linckia laevigata*). This shrimp feeds on starfish, turning them upside down and keeping them as a living larder. Pairs of shrimps may co-operate to overturn starfish far larger than themselves.

The foundations of almost all the invertebrate lineages that exist today were laid during this period, as well as many more that have not survived. *Cloudina* was soon joined by other species capable of building chalky reefs from their skeletons, like the vase-shaped sponge-like Archaeocyatha. In the half billion years since the Cambrian, the architects of 'coral reefs' have come and gone in cycles, each cycle ended by a mass extinction event, each new one begun with a spurt of evolution in the aftermath of disaster as new forms expanded into vacant niches. The fourth and present cycle arose from the aftermath of the greatest mass extinction of all. The end-Permian mass extinction began 252 million years ago and over a couple of million years wiped out an estimated 95 per cent of life on Earth.

RIGHT Jellyfish blizzard in Jellyfish Lake, Palau. This and several other lakes in Palau have been isolated from the sea for around 20,000 years. Their water is refreshed by seepage through cracked limestone rock, but the jellyfish have been trapped for millennia. Each of the five different jellyfish lakes supports a different subspecies of jellyfish (*Mastigias c.f. papua*), present in bewildering abundance, isolation having taken evolution in different directions.

All five mass extinction events in Earth history are associated with a massive release of carbon dioxide into the atmosphere which led to runaway global warming. During the end-Permian extinction, ocean temperatures in the tropics topped 40°C. But temperature rise was not the only problem for marine life and arguably not the worst. When carbon dioxide dissolves in water it produces carbonic acid, which in turn reduces availability of dissolved carbonate, the key building block of chalky shells and skeletons. Suddenly, great assets became liabilities and at each mass extinction, success stories of evolution were stopped in their tracks.

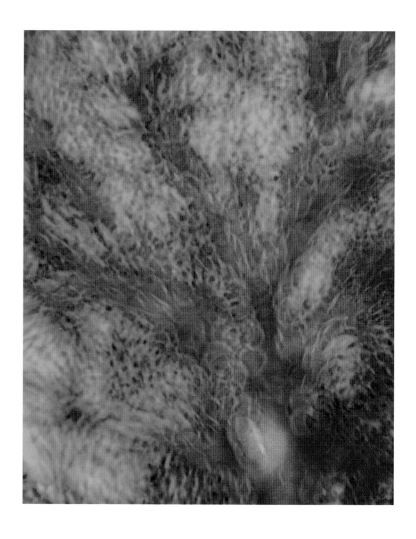

ABOVE Impressionist seafan. **RIGHT** Sea lily abstract.

FAR LEFT A Lion's Mane Jellyfish (*Cyanea capillata*) with feeding tentacles spread. Few creatures bother to eat jellyfish because the caloric rewards are so sparse. One exception is the giant Leatherback Turtle (*Dermochelys coriacea*) which migrates to high latitudes specifically to find high concentrations of jellies in the productive spring and summer plankton blooms. Despite their huge size, one study showed that Leatherbacks can eat three-quarters of their weight in jelly every day, amounting to hundreds of jellyfish.

LEFT Photo montage of different species of nudibranchs, or sea slugs, from Gulen in Norway. Sea slugs split from the shelled molluscs around 350 million years ago. They still grow a shell as larvae drifting in the plankton, but shed it when they metamorphose into juveniles and settle on to the seabed to live out the rest of their lives. Sea slugs eat creatures others shun, loading their bodies with the chemical and biological deterrents manufactured by their prey. Their brilliant colours warn would-be predators of their distasteful and toxic flesh.

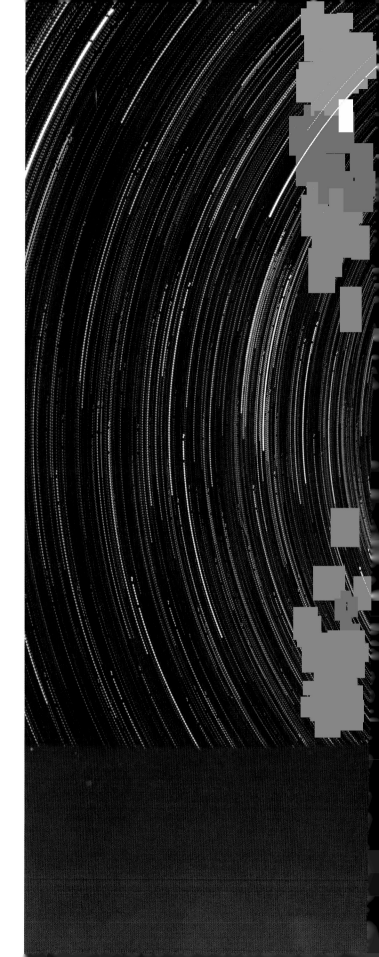

There are long gaps in the geological record after every mass extinction when 'coral reefs' disappear. But chalky shells and skeletons are such a benefit to those that possess them, that when ocean conditions returned to normal, new creatures proliferated to fill the gaps left by those that went extinct. A few tens of millions of years after the end of the Permian, a new kind of coral got together with a new kind of microscopic algae called zooxanthellae. Their extraordinary partnership has formed the basis of coral reef building ever since. Zooxanthellae live in coral tissues giving corals the ability to photosynthesise like plants, enabling them to grow fast and build reefs of enormous solidity. They survived the mass extinction 66 million years ago that killed the dinosaurs, and by chance, we find ourselves now in the heyday of these scleractinian corals. Since the last Ice Age ended, the world has seen the greatest surge of reef growth for 200 million years. That could be about to end and it's all down to us.

RIGHT Celestial sea slug *Flabellina pellucida*. The name 'sea slug' doesn't really do justice to the diaphanous grace of this creature. Nor really does its alternative name – nudibranch – which means 'naked gill'. This photograph digitally stacks 80 photographs on top of one another to add star trails in the Norwegian night sky.

Since the beginning of the Industrial Revolution in 1750, the oceans have seen a 30 per cent rise in acidity due to carbon dioxide emissions. If we continue to burn fossil fuels this way, acidity will rise 150 per cent by 2100, an increase ten times faster than in the end-Permian mass extinction. We are entering uncharted waters. Lab experiments and geological history both point in the same direction: life will get tough once again for creatures that build chalky shells and skeletons. Some foresee the end of coral reef building as we know it as early as 2050. Others are less certain. One likely outcome of our human-led transformation of the sea is a complete redefinition of what it takes to be successful. Species with chalky structures will probably suffer, whereas those without may prosper.

ABOVE There isn't much cover on the muddy seabed of the Lembeh Strait in north Sulawesi. This Veined Coconut Octopus (*Amphioctopus marginatus*) has adopted a marvellous method of protection. In a rare example of tool-use by an invertebrate, it drags two halves of a discarded coconut about with it, stilt-walking on tiptoes, and assembles them into a protective den when needed. This behaviour is not a one-off but has been seen repeatedly.

RIGHT A pair of courting Mimic Octopuses (*Thaumoctopus mimicus*). The smaller male rides on top of the female as he tries to place his sperm sac inside her mantle with his arm. Mimic Octopus is so-called for its ability to imitate venomous reef creatures by adopting their shapes and colour patterns. It's a nice idea, but Alex is not so sure. Having watched these animals for many hours, he thinks we may be seeing things that are not really there, like shapes in passing clouds.

LEFT A pair of Reef Octopuses (*Octopus cyanea*), probably two males, fighting in the Red Sea. Individuals of this species have been known to 'strangle' another to death by wrapping a tentacle tightly around the inhalant and exhalent openings of the mantle. In one case a female did this to a male with which she was mating, and then ate him!

Squid and octopus did well after the end-Permian mass extinction when the oceans were more acidic and low in oxygen and may prosper again in the coming century. But a ragtag group of shell-less invertebrates could do best of all, the gelatinous zooplankton. Animals like jellyfish, salps and comb jellies have seen their stock rise in recent years, benefitting from an unparalleled combination of favourable omens: they lack chalky structures so can shrug off ocean acidification; many like warmth, so global warming is not a problem; they love lots of nutrients, so sewage and pollution from farm runoff are beneficial; and they thrive when their predators are overfished. Jellyfish outbreaks seem to be on the rise, so much so that scientists have dubbed their recent success "the rise of slime". We may, they say, be replaying geological history in reverse, pushing the oceans back to the pre-Cambrian days of *Cloudina*, when jellyfish ruled. Whatever happens, invertebrates of some description will take full advantage of the opportunities. ✳

RIGHT A hyperiid amphipod, a type of crustacean related to beach sandhoppers, hitchhikes on a sea gooseberry (ctenophore). These amphipods quit the seabed long ago in their evolutionary history and have become space travellers in the planktonic galaxy. They ride jellyfish asteroids through the void, feeding on them, mating and giving birth on these tiny worlds. Females brood their young in a belly pouch and when they come close enough to the right kind of jellyfish floating by, they deposit their young there and wish them luck before they disappear for ever on journeys of their own.

Seaweed cathedrals

Kelp forests line huge stretches of temperate sea coast, and for those of us who live in mid-latitude countries are one of the most familiar ocean ecosystems. However, few people think much of them, only seeing a slippery and impenetrable tangle of lank fronds slopping at the surface. Approached from underneath in full summer, kelp forests are grand spaces with towering, shadowed arbours of wafting blades shot through by shafts of sunlight that play across the seabed as light pours through cathedral windows. These are veritable forests, albeit made up of fast-growing seaweeds, but unlike forests on land they are seasonal. The 'trees' die back to stumps every year as winter storms rip asunder the fronded naves and galleries, throwing the ruins ashore to rot in glutinous drifts.

LEFT Giant Kelp (*Macrocystis pyrifera*) stems rise heavenwards like cathedral columns. The floor of this forest is paved with rock stained purple by blotches of encrusting coralline algae and emblazoned with Red Gorgonians (*Lophogorgia chilensis*). A female California Sheephead (*Semicossyphus pulcher*) passes by to the right, her belly bulging from her latest meal.

Although kelp forests are found in all oceans, they reach their zenith in the Pacific. There, species like Giant Kelp and Bull Kelp achieve extraordinary lengths of 40, 60, even 80 metres. Kelps love cool water rich in nutrients and a hard bottom in which to root, so seaweed forests develop best on rock-bound, fog-shrouded coasts. When spring days lengthen, there are stirrings in the gnarled and knotted stumps that clasp rock slabs and press deep into their crevices. Then the plants race upwards, extending by up to half a metre every day to form a canopy that closes over by early summer, shrouding the understory in muted browns and greens. There is a reverential feel, wending your way among stems that rise toward the light as colossal vegetable columns, thrusting upwards from a lush understory of fleshy and coralline seaweeds that paint the rocks red, ochre, pink and orange.

ABOVE Hooded Sea Slugs (*Melibe leonina*) filter feed attached to a stem of Bull Kelp (*Nereocystis luetkeana*). Although they sieve food particles from the water while stationary, they can crawl and swim. The touch of a single starfish tube foot, signalling the arrival of a predator, is enough to make them take flight for a swimming burst a few minutes long.

RIGHT A Red Irish Lord (*Hemilepidotus hemilepidotus*) lurks in the Bull Kelp (*Nereocystis luetkeana*) forest of Browning Pass, Vancouver Island, with Quillback Rockfish (*Sebastes maliger*, top centre) and Copper Rockfish (*Sebastes caurinus*, left and right) behind. The most striking thing about Red Irish Lords, other than their magnificent colouration, is that they have a very rare means of parental care: up to four males co-operate to defend a nest of eggs laid by one or more females.

As the forest fills out the space between submarine rock and air, it draws in shoals of fish that hang in currents that pour between the stems, trailing blades like flapping banners. Fish nod and weave as they pluck plankton from the liquid conveyor that makes feeding almost effortless in this surfeited environment. Among the rocks below, burly rockfish, sculpins, eels and octopuses jostle for space, arguing with one another over the best hunting spots from which to surprise prey, or the best dens in which to crush and digest shellfish. Kelp fronds grow from the bottom up. As the season progresses, the older tips become ragged, shedding fragments that supply food for grazing urchins and abalones. When their protective slimy coat thins, the blades are colonised by legions of invertebrates, like sponges, bryozoans (sea moss), hydroids (sea nettles) and ascidians (sea squirts). They in turn attract nudibranchs (sea slugs), prawns, pycnogonids (sea spiders) and tiny fish.

RIGHT A young California Sealion (*Zalophus californianus*) caught napping in the canopy of a forest of Giant Kelp (*Macrocystis pyrifera*).

LEFT A quartet of California Spiny Lobsters (*Panulirus interruptus*) cram a rock ledge. Where rocky reefs have been protected from fishing, as here in the California Channel Islands, lobsters become far more common, filling every cranny. Their presence benefits the kelp, since they hunt sea urchins among other things. Without predators, sea urchins become too abundant for their own good, and that of kelp, mowing down seaweed until nothing is left but bare rock.

ABOVE Seaweeds have bendy stems so they can flex in the current or flop when the tide falls. Air-filled bladders like these on a Giant Kelp (*Macrocystis pyrifera*) plant buoy them up towards life-giving sunlight.

RIGHT A string of plump eggs girds the brood patch of this male Leafy Seadragon (*Phycodurus eques*). Seahorses and seadragons are in a minority of animals where fathers rather than mothers carry their unborn young.

Remarking on the occupation of the California Channel Islands some 12,000 years ago, very early in the human colonisation of the New World, archaeologists contend that kelp forests afforded colonists a highway from Asia to the Americas. Maritime skills honed on the kelp-slathered coasts of Japan and Kamchatka would have enabled people to follow the kelp trail around the icy Aleutian Islands, sustained by the bounty of the sea all the way to California and Mexico. Although boats were slight at the time, some little more than rafts made of reed bundles, kelp plants damped waves and so created a calm corridor to navigate between sea and land. The prolific maritime fertility of the Pacific Northwest enabled First Nations people to devote much time to creativity, developing sophisticated artistic cultures over thousands of years.

ABOVE A Giant Pacific Octopus (*Enteroctopus dofleini*) hunts in the open in Canada. In a poignant final act in their brief three to five year lives, females retreat into a den to look after their eggs. For six months or so each female protects her eggs from predators and wafts them tenderly with water to oxygenate them and keep them clean, all the while slowly starving. After the young hatch, the mother withers away and dies.

RIGHT A Grunt Sculpin (*Rhamphocottus richardsonii*) creeps through soft corals and sponges on the tips of its finger-like pectoral fins. This north Pacific fish has evolved to resemble the Giant Acorn Barnacle (*Balanus nubilis*), reaching about the same size of five to eight centimetres long. It lives in disused barnacle shells, or sometimes discarded bottles and cans. When ready to mate, a female chases males until she corners one in a crevice or hole, then lays her eggs and waits until he fertilises them before leaving him on guard. When the eggs are close to hatching, the male takes them into his mouth, swims into open water and spits them out, breaking the shells so the larvae can swim away.

LEFT Black Rockfish (*Sebastes melanops*) throng among the chunky stems of Bull Kelp (*Nereocystis luetkeana*) on Canada's Pacific coast. The extraordinary productivity of fish, shellfish, Sea Otters (*Enhydra lutris*) and other life in kelp forests has led some archaeologists to suggest that the Pacific Rim formed a 'kelp highway' on which maritime people colonised North America from Asia at the end of the last Ice Age.

ABOVE A Draughtsboard Shark, also known as the Australian Swellshark (*Cephaloscyllium laticeps*), in a Giant Kelp (*Macrocystis pyrifera*) forest in Fortescue Bay, Tasmania. The swellshark can gulp water and grow to twice its normal size, in order to wedge itself into rocky crevices and avoid being dragged out and eaten by larger sharks or seals.

Kelp forests provide cover for hunters and hunted. Killer Whales conceal themselves close to kelp along the west coast of North America so they can spring surprise attacks on Grey Whale calves making their first migration north to Arctic feeding grounds. Human whalers used the forests for the same purpose, waiting in their boats among the weeds until a migrating whale passed within reach. Sealions and fur seals hunt in them and escape sharks among their tangled blades.

While life in these forests bursts forth with an exuberance hard to find anywhere else, the apparent regularity of the seasonal transformation from rock to forest and back is surprisingly tenuous. It was in kelp forests that one of the first demonstrations was made of how the structures of ecosystems depend on close networks of interconnection among their inhabitants. On Pacific coasts of North America, Sea Otters were hunted from the 18th century, leading to their near complete extinction, which led in turn to disappearance of much of the kelp. The connection is not immediately obvious, but is easily explained. Kelp plants were overgrazed when a superabundance of urchins and abalones developed as a consequence of release from their otter predators. The same consequences play out in places that lack Sea Otters, like Australia and New Zealand, due to overfishing of fish and lobsters that eat urchins and abalones.

LEFT A school of Blacksmiths (*Chromis punctipinnis*) intercepts incoming plankton streaming through towering columns of Giant Kelp (*Macrocystis pyrifera*).

TOP LEFT Ochre Sea-stars (*Pisaster ochraceus*) clamber over an anemone-clad slope in a north Pacific kelp forest. In the 1960s these starfish, which can grow half a metre across, led to a new understanding of how some species play keystone roles in their ecosystems. When the starfish was removed from areas of kelp forest, the barnacles and mussels it preys upon soon came to dominate space, elbowing out kelp and the rich assembly of species that live among its fronds and holdfasts. Similar cases in which predators play lynchpin parts in structuring the habitat around them have now been documented from all over the world.

TOP RIGHT A Giant Spider Crab (*Leptomithrax gaimardii*) surveys the camera warily from a *Sargassum* plant. This south Australian native gathers in Port Philip Bay every winter in astonishing numbers to moult their shells and swap them for a larger one that has grown underneath. The new shell is soft before it swells and hardens so they are vulnerable to predators. The mass moult is possibly a tactic to overwhelm predators, ensuring that more crabs survive.

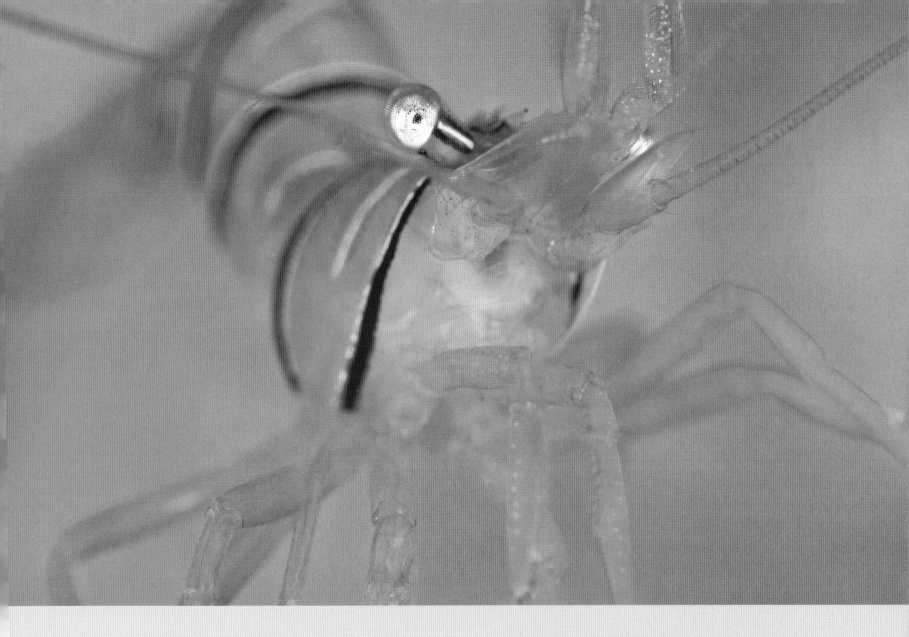

ABOVE A small Candy-striped Shrimp (*Lebbeus grandimanus*) nestles in the plush protective embrace of its host, pink sea anemone.

This cascade of unintended consequences helped cause the extinction of Steller's Sea Cow from the northern Pacific in 1768. This placid ten-metre beast was discovered and named by the naturalist Georg Steller when his boat was shipwrecked in 1741 on the return journey to Russia after discovering Alaska. At the time of the wreck thousands of sea cows lived around Bering and Copper Islands not far from the Russian mainland. The sailors escaped, eventually, and brought back with them hundreds of valuable Sea Otter pelts, triggering a hunting gold rush to the islands. With decline of the otters, the kelp dwindled and the last Steller's Sea Cow starved to death just 27 years after the species was discovered.

With a little help and lots of protection, Sea Otters have re-established themselves from Alaska to California, helping kelp dominate once again where it had all but vanished. Marine protected areas have rebuilt depleted predator populations to achieve the same miracle of regeneration in Tasmania, New Zealand, southern California, Mexico and Chile. ✳

RIGHT A male Weedy Seadragon (*Phyllopteryx taeniolatus*) carrying eggs wafts through the shady glade of a kelp forest in Tasmania, Australia. After two weeks of courtship and increasing intimacy, the female lays her eggs in sticky strings on to the male's brood patch. Immediately afterwards, he releases his sperm and swims in tight circles to fertilise them.

FAR LEFT Harbour Seals (*Phoca vitulina*) thrive on the lavish smorgasbord of fish and shellfish found in kelp forests. This one is pictured in California, but the seal is one of the most widespread in the world, occurring around all of the northern hemisphere's cooler coasts. The Pacific ones look rather different to those around European coasts, somewhat as Californians do to people in Europe!

LEFT Leafy Seadragon (*Phycodurus eques*) selfie, Yorke Peninsula, South Australia (well, not quite!).

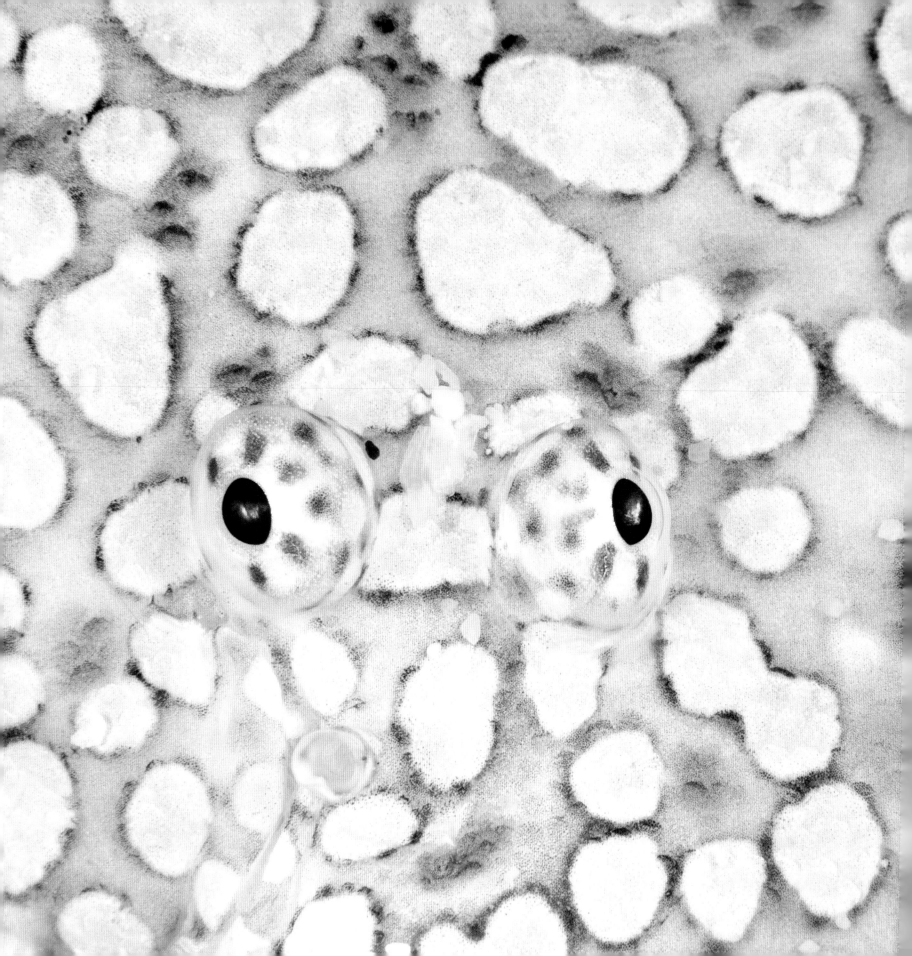

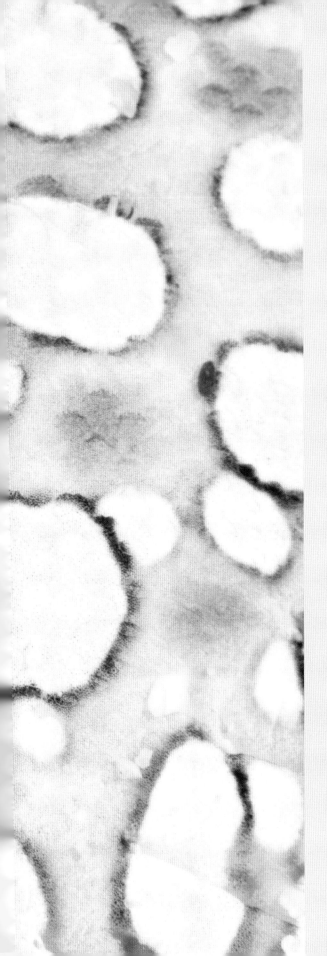

The nature of beauty

Fish are among the most striking life forms in the sea and are widely loved, so much so they have been given starring roles in movies. Their beauty and diversity of form are extraordinary: there is a breathtaking variety of ways to be a fish. What is it about fish that draws us to them? A study at the National Marine Aquarium in Plymouth, UK, found that the well-known soothing effects of aquaria were enhanced when there were more fish of a greater variety in the tank. Visitors lingered longer beside a tank when it had more fish, and within a few minutes of starting to contemplate them, experimental subjects connected to monitors experienced reduced heart rate and blood pressure.

LEFT There is little as changeable as a sole, here a Frill-mouth Sole (*Brachirus heterolepis*) that has matched its spot pattern to a gravelled seabed at Saonek Island, West Papua. Seen from further away, soles look like flattened teardrops.

That we find the diversity and abundance of life appealing probably comes down to our long evolutionary intimacy with nature. Wild creatures and places attract us on a deep subconscious level. Beauty is a very difficult idea to pin down, though. Plato thought beauty was a universal value, along with goodness, truth and justice. But while we may all experience the powerful emotions that beauty provokes, we know that what one person finds beautiful another derides, especially in the world of art. But there do seem to be common factors. People shown photographs of others find those with more symmetrical faces and smoother skin more attractive. This preference is so deeply embedded that it works across races. Asians shown photos of Caucasian people find the same ones attractive as Caucasians do and vice versa. Biologists have related symmetry to genetic fitness and smooth skin of course means youth and health. Neurologists have traced the experience of beauty to structures in the brain associated with sensory appraisal, in other words deciding whether something is good or bad for us, be it a food, a potential mate or a predator. Such scientific explanations of a characteristic that sets us apart from the rest of the animal world – our sense of aesthetic appreciation – are deeply anathema to some.

LEFT A male Tompot Blenny (*Parablennius gattorugine*), uses a discarded tube as a nest beneath Swanage Pier in the English Channel. His extravagant colours attract females to lay their eggs in his nest. He defends the eggs against all comers – hungry fish, crabs and starfish, even the photographer – for several days until they hatch into larvae that swim off into open water, never to be seen again by the parent.

BELOW For blennies, holes mean safety from predators. In a world where natural habitats are under pressure, rubbish can offer a welcome alternative, as for this Brown Sabretooth (*Petroscirtes lupus*) in Sydney, Australia.

What makes one fish beautiful and another not? Many fish are drab and simply want to disappear into their background. But good camouflage in itself is not always an impediment to our appreciation. Some fish are so extravagantly good at concealment, so ridiculously lavish in their mimicry, that we marvel at them. Frogfishes mimic sponges or rocks or invertebrate-covered reef so exquisitely that they are beautiful. Seahorses and their relatives are undeniably cute, but not all people think them beautiful, apart perhaps from those species that have so perfected their camouflage they blend in completely: like the candy-coloured pygmy seahorses that inhabit seafans, or the Weedy Seadragons that by means of colour and behaviour seem as much like plants as the kelps they swim among.

RIGHT There is nothing more arresting about a coral reef than the halo of fish around it, and among those fish there are few more striking than Lyretail Anthias (*Pseudanthias squamipinnis*). Although their shoals are in constant motion, plucking plankton from the current and pulsing back and forth between shelter and open water as predators pass, they have a hidden structure. Within them, slightly darker males defend and control harems of females. This school is mixed with a group of Waite's Splitfin (*Luzonichthys waitei*), which are purple-blue below with orange backs.

FAR LEFT A Bluestreak Cleaner Wrasse (*Labroides dimidiatus*) peeps from within the gill cavity of a Blue-spotted Pufferfish (*Arothron caeruleopunctatus*) in Thailand. Cleaner Wrasse are in high demand on coral reefs for their one-stop parasite and dead tissue removal service. They attract clients to cleaner stations with a characteristic head-nodding welcome, often delving fearlessly into the throats of predators ... and always emerging unscathed.

LEFT A Spotted Moray (*Gymnothorax melanospilos*) seen close up in Baa Atoll, Maldives.

Fear and beauty are intertwined in our evolutionary heritage. Creatures that are feared are often considered ugly, like spiders or snakes, and big, sharp teeth don't do much to improve our appreciation of a face. On the other hand, animals that are docile are often admired, especially if they are edible or friendly. It is almost impossible not to imbue fish with human personalities based on their looks. Frogfish and anglerfish look grumpy with their downturned mouths, an impression strengthened by their habitat choices, living among rubble or in muddy pockets in otherwise beautiful surroundings. The impassive faces of sharks or jacks seem cold and predatory; moray eels look sinister; parrotfish are goofy; groupers are surly with their fat-lipped scowls; blennies are comical with their silly antler hats.

RIGHT Two male Long-tail Dragonets (*Callionymus neptunius*) scuffle on a reef in Negros Island, Philippines. Fighting is a way of life for these fish, determining who gets to mate. Males defend patches of territory from others to secure access to a harem of females, so the fighting can be intense with biting and strident display of coloured fins. Bigger males with larger territories get more matings. During the peak season, fit males spawn with several different females every day.

The brightly coloured fish of tropical seas are at the opposite pole from drab. Most people find their dazzling colour contrasts instantly appealing, like black and yellow or blue and orange. Some fish have iridescent patterns that seem to glow, like the electric blues of a neon sign. But what do these colours and patterns mean to the fish? To understand that, we need to see like a fish and think like one too. Schooling fish are often striped longitudinally, a pattern that seems to signal peaceful co-operation in the fish world, whereas vertical bars are more often associated with aggression and territoriality.

PREVIOUS PAGES In highly productive waters like those of the Gulf of California, fish can reach such dizzying abundance that they form moving walls. This one is tiled with the silver bodies of Green Jacks (*Caranx caballos*).

CLOCKWISE FROM FAR LEFT Angelfish as art:
Blueface (*Pomacanthus xanthometopon*),
Emperor (*Pomacanthus imperator*) and
Bluegirdle (*Pomacanthus navarchus*).

The colours we perceive are not necessarily those the fish see. Some are sensitive to ultraviolet light, although UV is filtered quickly by water so these fish generally live near the surface. Some can see polarised light, which may reduce glare and enhance contrast. The vivid colours typical of flash photography underwater are often not representative of what a creature looks like in natural light. With increasing depth, water filters out red wavelengths first, then orange, then yellow, and so on until only blues and greys are left in the encroaching twilight 50 or 100 metres down. Red colours appear dark grey or even black by depths of 20 metres or more.

RIGHT A pair of Pygmy Seahorses (*Hippocampus bargibanti*), each no bigger than a peanut, wrestle in a seafan (*Muricella*) in Indonesia's Lembeh Strait. So exquisite is its mimicry, this species was only discovered and named after being inadvertently collected with its host seafan and placed in an aquarium. They usually occur in monogamous pairs, and sometimes many pairs live in the same seafan. Because they are so tiny, they give birth to no more than ten young at a time, far fewer than the hundreds produced by some of the larger seahorses.

LEFT With its cavernous maw full of teeth like glittering icicles, this Fangtooth Moray (*Enchelycore anatina*) from the Canary Islands can be pretty certain there will be no second chances for its prey.

Fish eyes are often fantastically pigmented, with flecks of gold, purple, yellow, red and 100 colours in between. It seems paradoxical to obscure vision this way, but the pigments act as filters, blocking certain wavelengths of light so the fish can see better. The pigments may reduce glare and increase image sharpness. Some fish can change the colours in their eyes to match changing light conditions. These functional explanations belie their undoubted aesthetic appeal. Some fish eyes are gorgeous seen close up. You can quickly lose yourself looking into a deep dark pool of a pupil flecked and spangled with gold dust, emeralds and rubies. In fact, seen close up, fish can become abstract works of art that evoke our sense of beauty on a new level. Separated from their 'fishiness', delicate coloured patterns of checks, zigzags, scales and fin-rays emerge to dazzle and mesmerise.

LEFT The mesmerising kaleidoscope within the eye of a Red Irish Lord (*Hemilepidotus hemilepidotus*). The flecks of pigment filter certain wavelengths of light, creating an internal shade that sharpens vision in the green drenched kelp forest interior.

BELOW This tiny Ghost Pipefish (*Solenostomus cyanopterus*) so perfectly mimics a drifting blade of seagrass, even down to the flecks of crud, that only the most experienced, and lucky, divers can spot them. Look carefully and you can see this one has just eaten a tiny fish which is revealed within its belly by the backlighting.

Do fish find each other attractive? Female fish often have to weigh up the pros and cons of mating with particular males. Perhaps they see subconsciously in certain males the beauty that we find in symmetrical, smooth-skinned people. When it comes to males, though, they seem depressingly indiscriminate – any girl will do, so long as she is mine. ✳

ABOVE Many fish have become masters of the art of concealment, whether it is to escape the notice of predators, or to fool potential prey. This Australian Tassled Anglerfish (*Rycherus filamentosus*) is so exquisitely camouflaged that most prey can surely have no sense of danger until finding themselves swallowed.

RIGHT The Hairy Frogfish (*Antennarius striatus*) is probably one of the most arresting examples of adaptive camouflage – if you notice it, that is. Looking like little more than a weed-covered rock, the frogfish uses a tiny worm-like lure on the end of a 'fishing rod' attached to its forehead to tempt prey close enough to swallow. This fish was photographed in typical mucky habitat of the Lembeh Strait, Sulawesi.

NEXT PAGES Ocean Sunfish (*Mola mola*), despite being the largest bony fish in the world (sharks and mantas have cartilaginous skeletons), are rarely glimpsed by divers. They are creatures of the open sea, living far offshore. Their name comes from the habit of lounging at the surface, where they must warm their muscles after bone-chilling dives hundreds of metres down in pursuit of their planktonic prey. In recent years, they have spread to higher latitudes, like the Gulf of Alaska, as the sea has warmed because of our greenhouse gas emissions.

Sea change

The first dawn rays colour the shallow reef edge in the Cayman Islands, contrasting it with the shadowy grey shapes deeper down where retreating night lingers. The first daytime fish make tentative sorties above the reef, keeping close to cover for now. Out in the open, three bolder fish move slowly across the face of the drop-off, light playing across fins spread like highly decorated Japanese fans. Their fins are joined near the base by translucent skin webs which separate at the tips into delicate stems, each striped and richly patterned in hues of rust, brick red and white. Two horns sprout from each forehead and a frill of fleshy knobs buds from each mouth. The explosion of colour and fin gives this animal one of its names, the 'fireworks fish'. That certainly seems preferable to 'chicken fish', which is another, perhaps from its similarity to showy cockerels. But this fish and its relatives are most often known as lionfish, which in view of their habits, is probably the most apt.

LEFT A Mountainous Star Coral (*Orbicella faveolata*) spawns at night. On a few nights of the year in late summer, corals all the way across the Caribbean release bundles of eggs glued together by sperm in a spectacular mass spawning. Nobody is sure how the corals know when to spawn. Mass spawning overwhelms predators which can eat only a tiny fraction of the bounty, so more of the young survive.

LEFT A Caribbean Reef Shark (*Carcharhinus perezi*) glides through a rich stand of gorgonians.

BELOW Portrait of a killer – the Red Lionfish (*Pterois volitans*). These lionfish are native to the Indian and Pacific Oceans but were introduced to the coast of Florida by aquarists in the early 1990s, either deliberately or inadvertently. They spread like wildfire and are now at home from the Gulf of Mexico to Bermuda and from Panama to Brazil. In their native seas, prey fish are wise and cautious but Atlantic fish are naïve and vulnerable. So the lionfish have become superpredators, gulping down prey until their bellies groan and growing fast to a size half as big again as the largest in the Pacific or Indian Ocean. They are driving down populations of native reef fish and there is widespread concern that they will change the Caribbean for ever.

They look like lords of this reef, wending their way among barrel sponges and seafans with a self-assurance that verges on recklessness. It is as if they know that nothing will trouble them, for they are by far the most dangerous fish here. What is most surprising about this scene is that these fish are imposters, swooping in only in 2008, like gangsters arrived to take over the town.

Red Lionfish, or *Pterois volitans* (the name means to hover or fly), come from the Indian and Pacific Oceans. They were first spotted in the Caribbean off the coast of Florida in 1992, no doubt the result of a well-meaning but deeply misguided release by an aquarist. This lionfish has become one of the most destructive alien invaders, as scientists call trouble-making non-native species, anywhere in the ocean world. In their native haunts, they are shy, usually uncommon and rarely reach more than 25 centimetres long. But in the Caribbean they seem to be pumped up on steroids, swaggering about the reef in mobs, gorging on native fish that have no idea how to deal with them, and reaching nearly twice the size.

LEFT A female Hairy Pipehorse (*Acentronura dendritica*) is a creature smaller than a computer pen drive that is thought to be transitional between pipefish and seahorses. Males look after the young but have open rather than sealed brood pouches where females lay their eggs. Despite its remarkable delicacy, the hairy pipehorse is found from the tropics all the way into Canada.

BELOW Schoolmaster Snapper (*Lutjanus apodus*) is one of the commonest Caribbean predators, sometimes forming huge shoals that rest by day and fan out in the evening to hunt through the night. Young Schoolmasters often live in mangrove forests, hiding among the dense framework of prop-roots to escape predators, while feeding mainly on tiny crabs and amphipods themselves. As they grow, fish move to reefs and take up residence there, switching diet to one with a lot more fish in it.

Lionfish are not the only thing that has changed here. Beneath its untroubled turquoise calm, the Caribbean Sea is in turmoil. Another possible invader, visible only by its effects, arrived in 1982 through the Panama Canal, having stowed away in ships' ballast water brought from the Pacific. This bug caused a disease that spread like wildfire, within a few years killing 99 per cent of the Caribbean's Needle-spined Sea Urchins. Loss of these grazing animals quickly led to a flush of seaweed that thickened rapidly and began to smother corals, especially in places where overfishing had depleted grazing parrotfish. Two other disease outbreaks romped through the region in the same decade, causing the calamitous loss of the two most important reef-building corals: Elkhorn and Staghorn Coral. Both are now listed as Critically Endangered on the global Red List of Threatened Species. One of these diseases was later identified as a human gut bacterium, no doubt introduced with sewage effluent.

ABOVE With so many predators about on a reef, creatures go to extraordinary lengths to protect their young. Here a male Yellow-headed Jawfish (*Opistognathus aurifrons*) blows out the clutch of eggs he is incubating in his mouth to oxygenate them. He sleeps in a chamber at the bottom of a burrow he has dug himself and which he closes at night for safety.

RIGHT A group of Tarpon (*Megalops atlanticus*) hunt silversides (*Atherinidae*). Tarpon can reach more than two metres long and look primordial with their giant goggle eyes and huge mirror scales, each the diameter of an apple. Surprisingly, perhaps, they are relatives of the much smaller herring familiar from temperate seas.

LEFT A Diamond Blenny (*Malacoctenus boehlkei*), no bigger than a human finger, seen through a window in a seafan. This one rests among the tentacles of a Giant Sea Anemone (*Condylactis gigantea*). There are no native anemonefish in the Caribbean, but this fish and several related species use the same trick to avoid predators, moving unharmed among stinging tentacles that could be lethal to others.

RIGHT A Spotlight Goby (*Elacatinus louisae*) peers from its home inside a Yellow Tube Sponge (*Aplysina fistularis*). There are more than 20 species of neon gobies similar to this in the Caribbean, inhabiting sponges or living coral heads. Individual species vary slightly in colour pattern across their geographic ranges, making them hard to identify with confidence. Some species play a similar role to the Cleaner Wrasse (*Labroides dimidiatus*) of the Indian and Pacific Oceans, picking parasites and dead skin from bigger fish. Others, like this one, feed on worms that live in sponges.

FAR RIGHT A Secretary Blenny (*Acanthemblemaria maria*) peers from a hole in a Cayman Islands reef. This tiny fish slots neatly into holes bored by molluscs, sponges and worms, where it lives its whole life. It darts out in split-second forays to grab passing plankton or to snatch hurried bites of tiny crustaceans that live in the low seaweed fuzz that coats the reef.

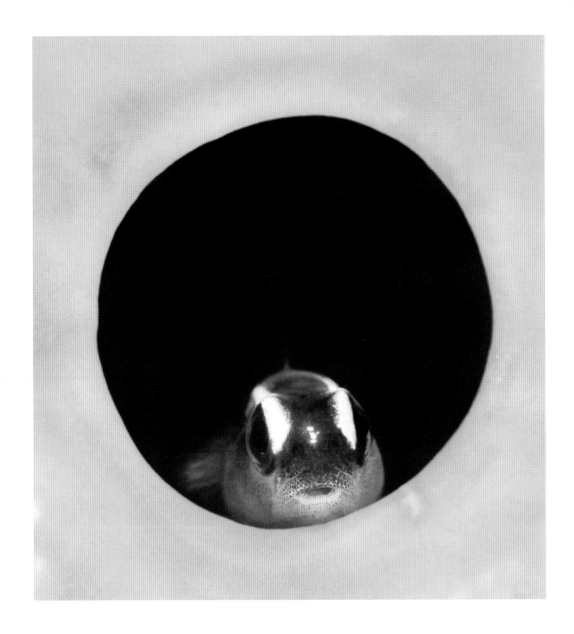

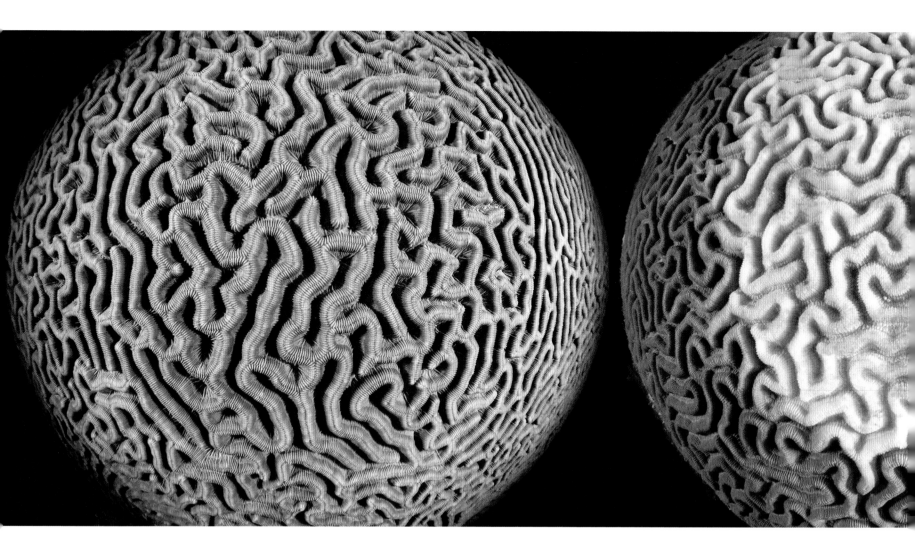

Other disease outbreaks have spread through seafans and corals like the Mountainous Star Coral, another key reef builder. In the space of a few decades, the Caribbean lost four-fifths of its stony corals, leaving many looking more like seaweed reefs than coral. Why has this region proven so fragile in the face of modern pressures? The answer has its roots in the origins of this sea three and a half million years ago when North and South America ploughed into one another, closing the Isthmus of Panama. Soon after, at least in geological terms, the Caribbean began to lose species in an extinction wave brought on by the reduction in habitat area and climate swings during Pleistocene Ice-ages. Today the Caribbean has just 61 coral species, eight per cent of the number found across the Indian and Pacific Oceans. Only four or five of them could be called serious reef builders, the rest playing bit parts in construction.

ABOVE Three Brain Corals (*Colpophyllia natans*) illustrate two of the biggest problems facing corals today. The left hand image shows a healthy colony. Its warm biscuit colour comes from microscopic algae called zooxanthellae that live within the coral tissue. The relationship between coral and plant is mutually beneficial, since the corals get much of their food from the plants, as well as some oxygen, while the plants get nutrients, carbon dioxide and protection. The central image shows a diseased Brain Coral, probably suffering from white band disease, a disorder that emerged in the 1980s. Tissue

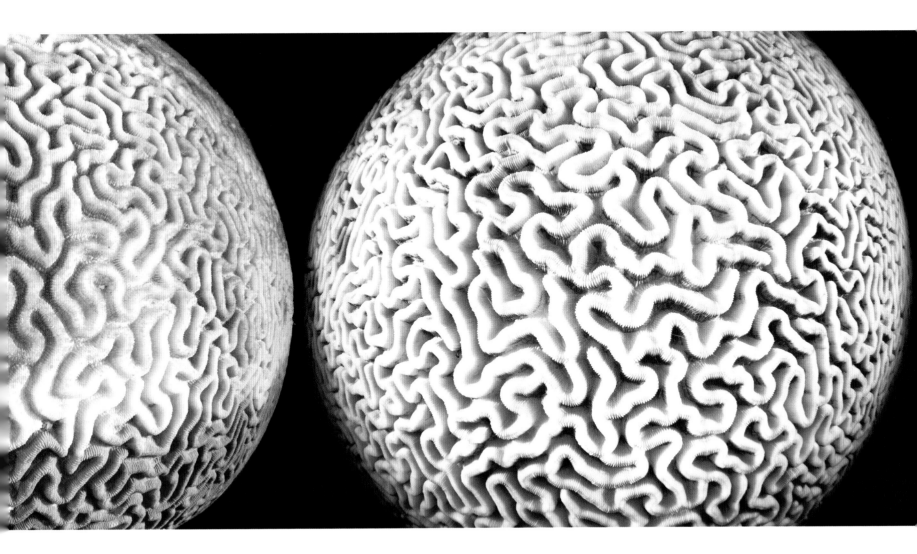

to the left and bottom of the photograph is still healthy, while inside the white band that shows the disease front moving across the colony, the flesh has been killed and the exposed skeleton is being colonised by green algae. A few weeks after this photograph was taken the whole coral would likely have been killed. Serious diseases now afflict all of the main reef building corals in the Caribbean, putting the future survival of reefs there in doubt. The right hand image shows a colony undergoing bleaching. Bleaching happens when the relationship between coral and plants breaks down, shifting

from benefit to cost when the coral is stressed. The coral either expels or kills the plants, leaving a layer of transparent tissue over bone white skeleton. Bleaching is most commonly caused by excessively warm water, which was the likely cause here. The coral is not dead, but if it stays like this for more than a month or two it will die because bleached corals are starving corals. If stresses are relieved soon enough, corals can regain new plants and recover.

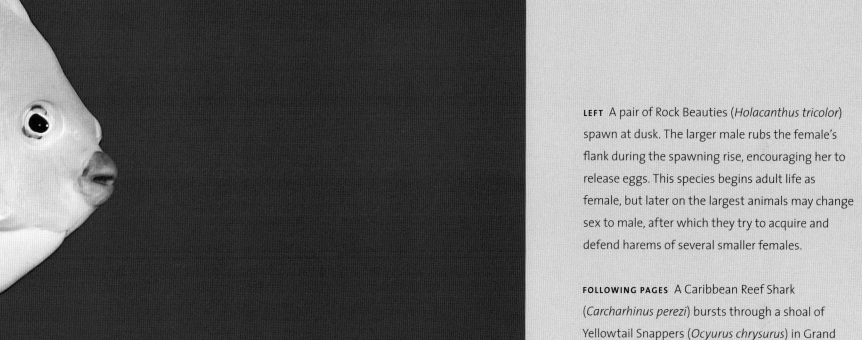

LEFT A pair of Rock Beauties (*Holacanthus tricolor*) spawn at dusk. The larger male rubs the female's flank during the spawning rise, encouraging her to release eggs. This species begins adult life as female, but later on the largest animals may change sex to male, after which they try to acquire and defend harems of several smaller females.

FOLLOWING PAGES A Caribbean Reef Shark (*Carcharhinus perezi*) bursts through a shoal of Yellowtail Snappers (*Ocyurus chrysurus*) in Grand Bahama. Reef sharks are among the first to dwindle when fishing intensifies and so few places in the Caribbean have healthy shark populations any longer. However, there are still plenty in the Bahamas and the country is leading a growing movement in the region, having declared its waters as a shark sanctuary. Hopefully, shark sanctuaries around other islands, like Bonaire, will pave the way for a comeback.

RIGHT A Queen Conch (*Strombus gigas*) in search of its next meal, a mix of seaweed and seagrass blades. This animal has been fished by the tens of millions over the centuries, having been exploited since prehistoric times. Since it is confined to shallow water within easy reach of divers, it has declined precipitously across its range. The Queen Conch was the first marine mollusc and the first major commercially fished species to be given protection under the Convention on International Trade in Endangered Species, having been listed in 1992. Improved fishery management has now secured its future across much of its range.

The comparative poverty of Caribbean reefs leaves them far more vulnerable to loss of species than the richer seas of the western Pacific and Indian Oceans, and probably more at risk to the addition of new ones. Low diversity (we are speaking in relative terms here, since the Caribbean is much richer than cooler seas) means more opportunity to lose animals and plants that play critical roles in their ecosystems, like reef-building corals. It also implies there are more opportunities for immigrants to gain a foothold and spread. Diseases and Red Lionfish are only early arrivals in what will likely prove to be an unstoppable flood of species from far and wide, swept in by globalisation. This small sea has become a crucible in which a new order is being blended from old and new ingredients. What changes lie ahead we can only guess at, but it is more than likely that the Caribbean will change as much again in the next 30 years as it has in the last.

BELOW A large stand of Staghorn Coral (*Acropora cervicornis*) on the North Wall of Grand Cayman Island. Before the 1980s, coral thickets like this dominated shallow reefs throughout the Caribbean. But an extraordinarily virulent disease swept through the region in the 1980s and wiped out almost all of them. Staghorn Coral is now critically endangered. The finger of blame points to us because the cause of disease has recently been identified as a human gut bacterium, probably released into the sea with sewage.

LEFT Backward-facing spines bristle inside a predator's mouth, the Caribbean Nassau Grouper (*Epinephelus striatus*), ensuring that when prey is grasped, it cannot escape.

ABOVE A pair of Indigo Hamlets (*Hypoplectrus indigo*) spawn at dusk. Hamlets are among the few vertebrate species that are true hermaphrodites, here the fish in the foreground is acting as a female, but this pair spawned several times during this evening, on each occasion exchanging sexual roles.

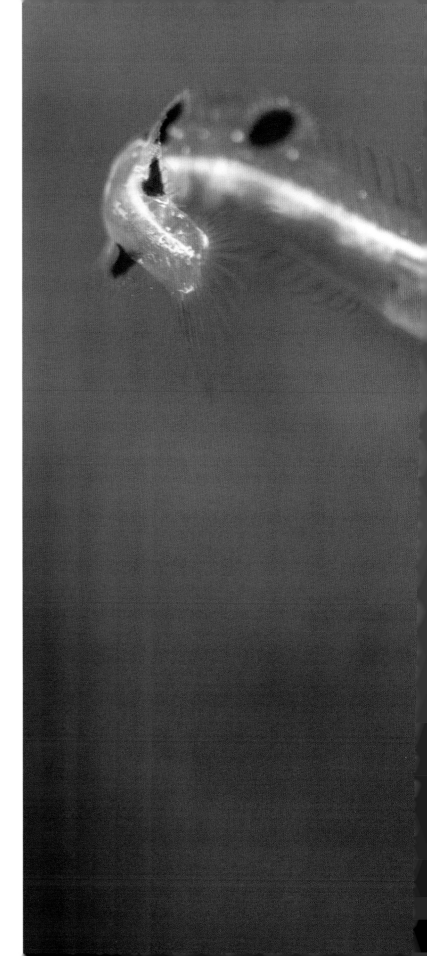

While the Caribbean has experienced momentous disruption, upending a period of stability that lasted 200,000 years, changes here are perhaps a harbinger of things to come in other places. The combined forces of global warming, sea level rise, ocean acidification, fishing, pollution and globalisation mean that nothing can be taken for granted any more. Change is the touchstone of our new world and will be for centuries to come. But although we may struggle to adjust, change does not have to mean disaster. The Caribbean is still an inspiring place of extraordinary loveliness, both above and below water, and will remain so if we give it a helping hand. As any good doctor will tell a patient struggling to cope: reduce stress, eat well to boost your immune system and take plenty of exercise to get fit. The same advice can be applied to the oceans to give them the best chance of seeing through the tough times ahead. But the stress-busting advice here would be for us to fish less, using less destructive methods, waste less, pollute less and protect more. We can no longer assume the oceans will take care of themselves. They need our help. ✳

RIGHT A tiny Arrow Blenny (*Lucayablennius zingaro*) yawns as it hovers above a coral reef, tail cocked and ready to shoot its streamlined body forward to catch its prey.

Desert ocean

Seen from space, the world's youngest and smallest ocean is a slender blue slash hemmed in between the desert ochres of Africa and Arabia. Viewed from the shore, the contrast between life underneath the Red Sea and that on its coasts could not be more surreal. For thousands of kilometres, the land is parched and barren. Up close, the dusty cliffs that rim the Red Sea are revealed as fossil coral reefs, petrified in their growth positions and thrown upwards by the geological upheaval that is creating this embryonic ocean. But beneath the surface, blocks of colour detach themselves from the dappling green and brown canvas that edges this sea to wander in an ever-changing impressionist portrait. Put on a diving mask and those roving colours become fish and the mottled canvas a seascape of living coral and seaweed.

LEFT There can be no more shocking contrast between exuberant life and its near absence than that between reef and desert in the Red Sea. Below water, life overflows from every nook and niche. Above sea level, the almost complete absence of water leaves desert rocks naked and unclothed by vegetation, producing a very different kind of beauty.

RIGHT Bohar Snappers (*Lutjanus bohar*) form a muscular bronzed wall of bodies as they gather to spawn off the Ras Mohammed headland, which forms the southern tip of Egypt's Sinai Peninsula. Snappers like this often gather in their thousands to spawn off reef promontories. Their eggs will be quickly carried offshore by strong currents, taking them to safety away from the scrum of hungry mouths that fronts the reef.

The difference between these juxtaposed worlds is water. Desert shores have little of it, the sea a never-ending supply, so life thrives in one realm and scarcely exists in the other. The shock of life underwater – of pulsing, surging, thronging, kaleidoscopic superabundance – is all the greater for its contrast with the world ashore. There is a puzzle here, though. The cerulean waters of the Red Sea are extraordinarily clear. Swim out over the shallow reef platform pretty much anywhere in the northern Red Sea or Gulf of Aqaba and you are confronted at the edge by a dizzying plummet into blue abyss. Clear water and precipitous verticality conspire to thrill and unnerve. Remarkable transparency implies an absence of impurities, which means the water contains very little by way of nutrients or plankton. Here is the paradox that baffled Charles Darwin: how do coral reefs sustain so much life in the midst of the low nutrient deserts of tropical seas?

RIGHT The moment of release in a Bohar Snapper (*Lutjanus bohar*) spawning aggregation. The white clouds are mixing sperm and eggs. These fish were photographed around the Pacific island of Palau. Bohar Snapper spawning has never yet been witnessed in the Red Sea.

LEFT Oceanic Whitetip Sharks (*Carcharhinus longimanus*) with their phalanxes of Pilotfish (*Naucrates ductor*) were once a familiar, everyday sight for oceangoing sailors. But their numbers plunged as fishers spread to the open sea, setting longlines studded with thousands of hooks to take tuna, swordfish and marlin. To begin with, sharks were unwelcome bycatch, almost worthless. But their fins are prized today for shark-fin soup, especially the large fins of this fish (*longimanus* means long arms), and fetch high prices in Asia.

It took 150 years to find an answer. Coral reefs, it turns out, are brilliant at capturing and holding nutrients, recycling them over and over through their complex web of life. The first thing that strikes most visitors to a reef is the halo of tiny fish that surrounds it. At the seaward edge where reef and ocean meet, so abundant are these miniature plankton pickers, it can feel like swimming through a psychedelic blizzard. They form a 'wall of mouths' that captures and concentrates sparse nutrients from open water. Hard corals, sponges, seafans, soft corals and dozens of other filter-feeding invertebrates carpet the reef, sucking in more plankton and particles of food recycled from other creatures. Reefs are full of grazers, predators and detritus feeders that swiftly pass nutrients from one to another. When predatory fish defecate, their poo rarely sinks far enough to hit the bottom; instead it is snapped up as a nutritious snack by any one of dozens of species. There is an epicurean hierarchy about fish poo. Some predators hardly have time to squeeze one out before a clutch of eager mouths is fighting over the spoils. Herbivore poo is rated only by detritus feeders, while the poo of the detritus feeders themselves is left for sea cucumbers and worms.

LEFT A globe of Bohar Snappers (*Lutjanus bohar*) hangs above the reef at Ras Mohammed ... waiting. When the right moment comes they will spawn together in a writhing frenzy of bodies, releasing their eggs as far from the reef and its predators as they dare. Sharks often take advantage of their preoccupation to slake their appetites on prime snapper fillet.

ABOVE A school of Hardyhead Silversides (*Atherinomorus lacunosus*) transforms an ordinary jetty into a bewitched woodland with a shimmering canopy of silvered leaves.

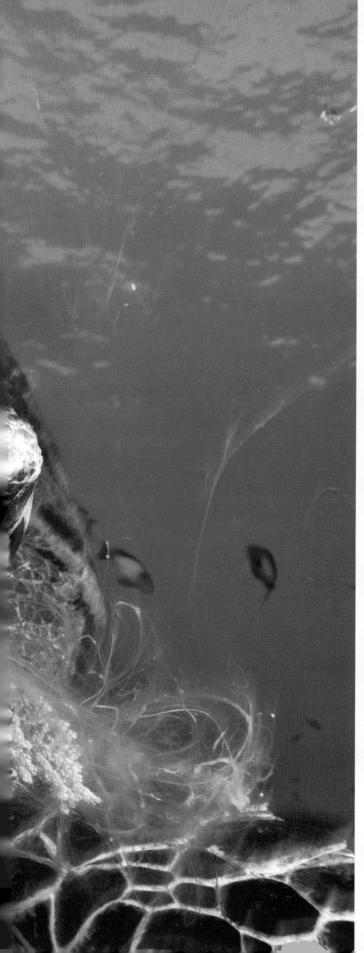

LEFT A grizzled Hawksbill Turtle (*Eretmochelys imbricata*) grasps a soft coral. Hawksbills eat animals that are shunned by most creatures because they are full of toxins to deter predators. Their flesh was distasteful to the sailors of old because of a build-up of these toxins so they were only eaten in extremis, but their shells were prized as 'tortoiseshell' so they too were slaughtered in millions. All sea turtles are considered endangered today but most species, including Hawksbills, are recovering following protection.

ABOVE This Spiny Spider Crab (*Achaeus spinosus*) on a soft coral scratches its head in comical perplexity, perhaps wondering what to make of the camera that is centimetres away.

Predation dominates every aspect of life on the reef. A fish is never more than a few moments of inattention away from death. Everything a crab, worm or snail does is fashioned around the twin desires to promote life and avoid death. Coral reefs turn the idea that predators are always less abundant than their prey upside-down. On pristine reefs that have never been fished there can be a far greater abundance of predators than prey by weight. Toothy morays lurk in caves, groupers group under ledges, snappers snooze in lazy shoals, while the steel flashes of pack hunting jacks, tuna and barracuda glint dangerously against the dark of open water. This inversion of common sense is possible because prey animals are small and short-lived and their populations turn over much faster than those of their predators. Prey and predators are like small and large cogs interlocking in a clock, spinning at different speeds. And unlike on land, the predators here are cold-blooded with much slower metabolism than mammalian mega-predators. They take less energy to sustain and can bide their time between meals.

RIGHT A huge Napoleon Wrasse (*Cheilinus undulatus*) in full-on threat display. This one was not upset by Alex, but by another male wrasse just behind his shoulder.

ABOVE A vast colony of the coral *Porites nodifera* dominates this coral bommie at Fury Shoal, Egypt. A couple of other coral species have gained a toehold, settling in spaces where patches of the big coral have died and then elbowing it aside as they grew. Corals compete for space by stinging each other or digesting their neighbours with filaments extended from within their guts. Some have more potent means of attack or defence than others, forming competitive hierarchies.

RIGHT A Scissortail Sergeant Major Damselfish (*Abudefduf sexfasciatus*) is the last of the daytime fish to seek its night-time roost as the sun sets over Egypt's Gubal Island in the mouth of the Gulf of Suez. Soon nocturnal fish will pour forth from crevices and caves and spread out in search of food. It is here at the seaward edge of Red Sea reefs – where they fall away into the deep water of open sea – that the richest life can be found.

LEFT A constellation of fish above the lip of a Red Sea coral reef.

ABOVE Blackfin Barracuda (*Sphyraena qenie*) circle off the wall at Shark Reef, Ras Mohammed, in Egypt. Barracudas were much feared by pioneering scuba divers, more for their predatory looks than their temperament. Attacks on people, although known, are very rare. This species is harmless to us.

The most fearsome predator of all in this world is not a shark or stonefish, but us. Throughout huge tracts of the global ocean, humans have become the dominant predators and overfishing has changed the face of reefs. Big-bodied and predatory fish suffer the heaviest losses, thanks to combination of slow growth, the high esteem in which we hold their flesh, and their bold habits. They dwindle quickly in the face of fishing, leaving behind resilient creatures with smaller bodies, faster growth and quickfire reproduction. You can tell how much a reef has been fished by the number of chunky predators it supports. Lots of sharks and large groupers means practically no fishing, while if there are few fish larger than a person's hand it means far too much.

LEFT Corals and seaweeds build reefs from limestone, which dissolves slowly in water to form caves. Reef caves like this one at St John's Reef in Egypt were probably formed by running fresh water when the sea level dropped 100 metres or more during the last Ice Age.

RIGHT Portrait of a male Scalefin Anthias (*Pseudanthias squamipinnis*). Some males control harems of female fish while others live in groups. Males with harems dance in the evening to entice their females to spawn. But they must be quick to take advantage of female interest since males without harems will dash in and release their sperm too, clouding the question of who sires the young.

FAR LEFT A male Bandcheek Wrasse (*Oxycheilinus digrammus*) spots its reflection in the camera. Males of this species are intensely competitive and one once launched repeated attacks on its reflection in Callum's mask, vigorously defying his efforts to swat it away.

LEFT A dainty female Coral Grouper (*Cephalopholis miniata*) flashes her pale, submissive colouration at a larger male. The male guards a harem of several females, mating with them frequently.

Nature is forgiving in these fruitful seas if you give her a chance. Protection from fishing by the establishment of marine parks soon leads to a resurgence in life. With robust enforcement, the abundance of predators can soar, rising five times or more in a decade. But it could take half a century to piece back together something that resembles a place that has never been fished. Reefs of the Egyptian Red Sea are at the forefront of such an effort. Beginning in 1983 with protection of the dramatic headland of Ras Mohammed at the southern tip of the Sinai Peninsula, Egypt has gone on to set up a string of marine parks covering hundreds of kilometres of coast. Today these waters are refuge for huge aggregations of spawning snappers and emperor fish, hosts of parrotfish, spiralling shoals of barracuda and jacks, grizzled turtles and elegant sharks. This protection makes the Red Sea one of the best places in the world to experience coral reefs in all their vivacity, spectacle and grandeur. ✳

ABOVE A male Red Sea Anemonefish (*Amphiprion bicinctus*) rubs a tentacle of a Magnificent Sea Anemone (*Heteractis magnifica*) over eggs freshly laid by the large female on the left. He covers the eggs in anemone mucous to protect them from further stings. Anemonefish have a mucus coat several times thicker than other fish to protect them from stings. They also rub themselves on the anemone to pick up its mucus as a disguise, making the anemone treat them as parts of itself rather than food or enemies, so they are not stung any more.

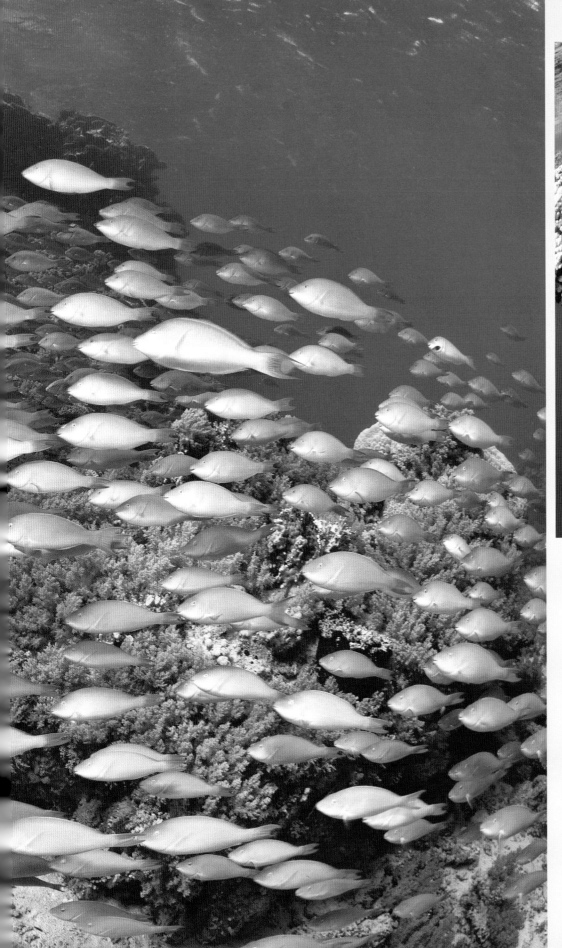

LEFT Longnose Parrotfish (*Hipposcarus harid*) swim purposefully in the late afternoon. Many reef fish make daily migrations to traditional rendezvous points, often around submerged headlands, where they spawn in frenzied rushes up and away from the reef. The schools grow in size along the way as more fish join. They make their way back later on in scattered groups when the orgy is over.

ABOVE The drop-off: a male Klunzinger's Wrasse (*Thalassoma klunzingeri*) flits above the vertiginous plummet where reef meets open sea.

Back from the brink

Seals are creatures of wild and wave-swept coasts, of remote rocks and beetling cliffs, uninhabited islands, deep caves and faraway ice. They haunt the edges of visibility, glimpsed basking on distant shoals or bobbing among kelp offshore. Their elusive habits gained them a mythical reputation in centuries past. The Selkies of Scottish and Faroese folklore took the form of seals in the water but shed their skins to become humans on land (where they often fell into doomed love affairs with land-bound people). Seals were imbued with magical properties. Romans carried pieces of Mediterranean Monk Seal skin to ward off thunderbolts and hailstorms. But our view of seals as remote and unapproachable is gradually changing today, because we are changing and therefore so too are seals.

LEFT A California Sealion (*Zalophus californianus*) near La Paz, in the Sea of Cortez. This sealion's stiff whiskers were much in demand in the 19th century to clean pipes in the opium dens of San Francisco.

Seals and sealions evolved from relatives of bears and dogs about 20 million years ago. Unlike whales and dolphins, they have not fully mastered the ocean and must leave it to give birth and nurture their young. Underwater they are graceful and lithe, frolicking with one another in fluid leaps and energetic somersaults. On land they are oppressed by gravity's pull on ungainly bulk, heaving themselves across beach and rock like enormous blubbery caterpillars. Evolution's flawed compromise between land and water makes them extremely vulnerable to predators when ashore, which is why most pup in wild places.

RIGHT Male Steller Sealions (*Eumetopias jubatus*) are burly, muscular and bold. Although it is hard to tell from the picture, this one charging past Alex at Race Rocks, Vancouver Island, is close to three metres long.

FAR RIGHT "This is my best side": a Grey Seal (*Halichoerus grypus*) poses fetchingly. It got its orange nose snuffling around in a rusty shipwreck!

By now it will be no surprise that the most dangerous predator seals faced was people. Humans have hunted seals for tens of thousands of years because these animals are enormously useful. Seals supplied a long list of commodities: waterproof hide, warm fur, floats for fishing nets, strong rope (made from skin cut in a long spiral), medicines, meat, oil for light, even ivory from Walrus. Cosquer Cave, which was once dry but today lies half submerged on the Mediterranean coast, contains paintings made 19,000 years ago of Mediterranean Monk Seals being hunted with spears. Male and female Grey Seals were engraved between 10,000 and 17,000 years ago on a piece of antler found in Montgaudier Cave in France's Dordogne. The seals are beautifully detailed and chase a salmon in characteristic belly up swimming position, showing that the artist knew them well.

RIGHT Shellfish are unforgiving foods, as this toothless old Grey Seal (*Halichoerus grypus*) demonstrates.

BOTH PICTURES A young California Sealion (*Zalophus californianus*) plays with a starfish. Pups pick up a variety of objects, like starfish, lumps of coral, shells, even feathers, and carry them to the surface, drop them and chase them to the bottom.

Over the centuries, human hunters drove seals to the edges of our world. Mediterranean Monk Seals are one of the most endangered today and perhaps the most evasive, giving birth deep in inaccessible caves. But it was not always so. Ancient texts from Homer's *Odyssey* onward suggest they were much more abundant in antiquity and bred on open beaches. But hunting took its toll and by 301 AD, a list of prices for goods and services published by Diocletian showed that a monk seal pelt cost 1,500 denarii, more than the price of a Lion or Leopard skin, and 15 times that of a bear. Clearly by this time the monk seal was a prized rarity. Accounts of where the seals could be found also shift with time as colonies of open coasts were hunted to disappearance. Medieval explorers later found new colonies of thousands of animals on the African coast, Canary Islands and Madeira, which they soon despatched by overhunting.

RIGHT Young California Sealions (*Zalophus californianus*) play in the early morning sun.

FAR LEFT Sea lions, with their scrawny bodies and indifferent fur, escaped the catastrophic 18th and 19th century human slaughter of fur seals with their rich pelts and Northern Elephant Seals with their blubbery bodies. Even so, like most seals and sea lions, they bred mainly on remote beaches or islands, far from people and other predators. Today, however, California Sea Lions (*Zalophus californianus*) seem to prefer places frequented by people, appearing quite at home in harbours and marinas, hauled out in noisy, smelly heaps of bodies.

LEFT An Australian Sealion (*Neophoca cinerea*) basks in the sun at Kangaroo Island, South Australia. All sealions have huge front flippers that they use like flapping wings to propel them underwater.

Seal hunting intensified in the 18th and 19th centuries as seafarers explored farther afield and their commodities were industrialised. Fur seal pelts were taken by the million from remote colonies on Pacific islands like Juan Fernandez and Sub-Antarctic Islands like South Georgia. Beaches of the South Shetland Islands were cleared of hundreds of thousands of seals within just three years of their discovery in 1819. When whalers ran short of quarry in the mid-19th century, they switched to pursuit of huge elephant seals throughout the Southern Ocean and into the North Pacific. Populations plummeted, island by island, beach by beach. Even well into the 20th century, Southern Elephant Seals were hunted alongside whales in South Georgia, ending up as margarine on European and American dinner tables.

RIGHT A bull California Sealion (*Zalophus californianus*) scatters baitfish. It is easier to see from his big head and shaggy mane, quite unlike the sleek elegance of females, how the name sea 'lion' was coined.

Unfettered hunting led to catastrophic losses, driving the Caribbean Monk Seal extinct and pushing many others to the brink. But loss of seals and falling demand for blubber and fur in the 20th century led first to a decline in hunting and later on active protection. These efforts have paid spectacular dividends. Guadelupe Fur Seals from coastal California and Mexico fell to a few dozen animals by the early 20th century, but have recovered to around 20,000 today. Northern Elephant Seals were down to their last hundred by 1890, but have bounced back to well over 200,000 now. The Juan Fernandez Fur Seal was long thought extinct until a few hundred animals were sighted in the 1960s. It has since made a comeback, reaching more than 30,000 by 2005 (although still a long way below the millions before exploitation). Among the eight species of fur seal, only the Galápagos Fur Seal is still in decline, although ironically it was never hunted much because being a tropical animal it has an indifferent pelt. Only a few seals are still hunted commercially today, like Canada's highly controversial hunt of new-born Harp Seals for their white fur.

LEFT There is an effortless grace about seals underwater that could not contrast more strongly with their clumsy lurching gait on land. Here a New Zealand Fur Seal (*Arctocephalus forsteri*) is caught on re-entry from a balletic surface leap.

BELOW A crèche of California Sealion pups (*Zalophus californianus*) practices their swimming skills in shallow water in the California Channel Islands. Young sealions are snack food for Killer Whales (*Orcinus orca*) and bite-sized dinners for Great White Sharks (*Carcharodon carcharias*), both of which are common here. They have to learn fast.

LEFT Familiar and hidden worlds are separated by the thin veil of the sea surface. A Grey Seal (*Halichoerus grypus*) takes a breath between dives into the underworld.

ABOVE Is this the seal equivalent of thumb-sucking? A large bull Grey Seal (*Halichoerus grypus*) sucks a kelp frond as he sleeps lodged between boulders at Lundy Island in England's Bristol Channel.

Decades of protection are changing behaviour. Fear of people is waning and animals are returning to breed on mainland beaches once more, like Grey Seals in the UK. Seals pop up and linger beside boats, kayaks and surfers. Young seals and sealions play with scuba divers like puppies, nipping their fins and pressing faces close to masks and cameras. So vigorous has seal recovery been that fishermen in some places now complain that the seals take their fish. But really, the problem is not seals but excessive fishing having depleted the fish that both seals and people depend on. The seal comeback shows that wildlife can thrive in a human-dominated world. But there is much more to do. We still have to find the right balance between fishing and protection for the world oceans. ✳

RIGHT A Harbour Seal (*Phoca vitulina*) investigates Alex's leg. Seals have many similarities with dogs, to which they are related, including their great sense of fun.

FAR RIGHT Seal selfie. Alex poses with a friendly young Grey Seal (*Halichoerus grypus*).

Index